Praise for *The Reindeer Chronicles*

"Thoughtful and thought-provoking, Judith Schwartz's world tour of environmental solutions shows how nature itself can heal the wounds we have inflicted on our planet. Compelling, fascinating, sometimes unexpectedly moving, this vitally important book is, above all, a springboard for hope and transformation."

—ISABELLA TREE, author of *Wilding*

"In *The Reindeer Chronicles*, Judith Schwartz proves, once again, that she is one of ecology's most indispensable writers. Like her last two books, *Cows Save the Planet* and *Water in Plain Sight*, her new work is an insightful, globe-trotting exploration of promising techniques for restoring our soil, water, agricultural systems, and wildlife. *The Reindeer Chronicles* is at once visionary and pragmatic—clear-eyed about the immense planetary challenges we face, yet unfailingly hopeful about our ability to forge a new relationship with nature. This book shows us what Aldo Leopold's land ethic looks like in the twenty-first century."

—BEN GOLDFARB, PEN America Literary Award-winning author of *Eager*

"This book shows us again and again, across the globe, the abundant future that is possible if we work *with* nature. Stunning stories of re-greening landscapes, restoring carbon and water cycles, and repairing weather. It is a balm and a guide, a wellspring of grounded hope."

—AYANA ELIZABETH JOHNSON, PhD, founder of Urban Ocean Lab and of Ocean Collectiv

"Judith Schwartz unlocks yet one more door in our minds about what's possible when we work with nature's cycles rather than try to push her around. Through this book and her prior ones, you can practically see, taste, and smell a healing earth that includes humans as stewards, not ravaging locusts. If you want practical hope, this is it. If you want a place to dig in and make change, regeneration is the key. These are stories of people who work both intimately and at scale—and with love—to restore life to the land we all walk on, our beautiful home, the earth."

—VICKI ROBIN, coauthor of *Your Money or Your Life* and author of *Blessing the Hands that Feed Us*

"A tale of people restoring nature and their communities. These deeply optimistic dispatches from around the world show us that the key to restoring land is how we see it—the change begins in us."

—DAVID R. MONTGOMERY, author of *Growing a Revolution*

"As the regenerative agriculture movement grows worldwide, Judith Schwartz has emerged as a leading tracker and interpreter of its progress, challenges, and wins. The value of Schwartz's multifaceted work and engaging first-person style is that a broader and deeper canvas emerges.

"Schwartz's descriptions and analyses are not rosy-eyed, but instead comprise a balanced, warts-and-all approach mixed with extraordinary tales of transformation of vast and small ecosystems, landscapes and farms, societies and communities; of food systems; and of human physical and mental health. As she says, 'earth repair is a participatory sport,' and 'restoration can begin anywhere.'

"This is an excellent read for expert and newcomer alike, and an important contribution to a growing canon now offering some of the very best solutions to the onrushing Anthropocene crisis."

—CHARLES MASSY, author of *Call of the Reed Warbler*

"These are times that call for us to reimagine everything. That imaginative capacity depends on the stories, the possibilities, the experiences we have in our memory and our ability to reassemble them in new and unique ways. If you want to be part of that reimagining, you need the beautiful, patient, humbling stories in these pages. Their implications are staggering, and also suggest that sometimes we save the world by doing less rather than more. Do your imagination, your activism, your sense of what's possible a favor, and swim in this book."

—ROB HOPKINS, author of *From What Is to What If*

The Reindeer Chronicles

Also by Judith D. Schwartz

Water in Plain Sight:
Hope for a Thirsty World

Cows Save the Planet:
And Other Improbable Ways
of Restoring Soil to Heal the Earth

The Reindeer Chronicles

And Other Inspiring Stories
of Working with Nature
to Heal the Earth

JUDITH D. SCHWARTZ

Chelsea Green Publishing
White River Junction, Vermont
London, UK

Cover image of reindeer migrating through the Kildal Valley (Finnmark, N.Norway) to summer pastures taken by Bryan and Cherry Alexander.

Project Manager: Sarah Kovach
Developmental Editor: Brianne Goodspeed
Copy Editor: Laura Jorstad
Proofreader: Natalie Jones
Indexer: Shana Milkie
Designer: Melissa Jacobson
Page Composition: Abrah Griggs

Printed in Canada.
First printing July 2020.
10 9 8 7 6 5 4 3 2 1   20 21 22 23 24

Our Commitment to Green Publishing
Chelsea Green sees publishing as a tool for cultural change and ecological stewardship. We strive to align our book manufacturing practices with our editorial mission and to reduce the impact of our business enterprise in the environment. We print our books and catalogs on chlorine-free recycled paper, using vegetable-based inks whenever possible. This book may cost slightly more because it was printed on paper that contains recycled fiber, and we hope you'll agree that it's worth it. *The Reindeer Chronicles* was printed on paper supplied by Marquis that is made of recycled materials and other controlled sources.

Library of Congress Cataloging-in-Publication Data
Names: Schwartz, Judith D., author.
Title: The reindeer chronicles : and other inspiring stories of working
   with nature to heal the earth / Judith D. Schwartz.
Description: White River Junction, VT : Chelsea Green Publishing, 2020. | Includes bibliographical references and index.
Identifiers: LCCN 2020019740 (print) | LCCN 2020019741 (ebook) |
   ISBN 9781603588652 (paperback) | ISBN 9781603588669 (ebook)
Subjects: LCSH: Restoration ecology. | Reclamation of land.
Classification: LCC QH541.15.R45 S393 2020 (print) | LCC QH541.15.R45 (ebook) | DDC
   333.73/153—dc23
LC record available at https://lccn.loc.gov/2020019740
LC ebook record available at https://lccn.loc.gov/2020019741

Chelsea Green Publishing
85 North Main Street, Suite 120
White River Junction, VT 05001
(802) 295-6300

Somerset House
London, UK

www.chelseagreen.com

This book is dedicated to the memory of my father,
Alvin D. Schwartz, MD (1932–2020),
who devoted his life to healing.

Contents

Introduction

It's time to rebuild what has been lost.

—United Nations Decade on
Ecosystem Restoration, 2021–2030

The biologist and pioneer of ecological design John Todd once wrote, "That which has been damaged can be healed."[1] I remember when this notion became personal for me. It was an evening in early June, our first dinner on the deck. I looked down over the meadow, across overlapping layers of green: the flared crowns of trees, grasses flecked white with wild strawberry, and the garden, where stalks of garlic rose from their mulchy beds. Green all the way to the blackberry thicket at the bottom of the hill: a prickly moat guarding the forest. From the stillness I heard our state bird, the hermit thrush. Its fluted call, heard only this time of year, always evokes for me the possibility of earthly perfection—joy captured in sound.

Southern Vermont puts on a good show in late spring. Still, thanks to what I understood from writing about soil and land health, I knew our small acreage could be more lush and productive. Our old apple trees—like most nineteenth-century Vermonters, the farmers who once lived here grew apples for cider—are being crowded out by brush. The hillside slope is overgrown. And after migrating tentatively year after year, the black raspberry vines have largely disappeared, as have the thimbleberries (the thimble-shaped red berries sit on your tongue and dissolve into pith).

People often ask me what they can do personally to help the environment, and I always say to start where they are. My husband, Tony, and I have mostly managed the land through benign neglect. Now I wondered: Which plants and in what numbers would help the land be vibrant and sustain diversity? What would "peak backyard" look like? I mentally conjured up photographs of restoration projects I'd followed—food forests dripping with fruit, fields of thigh-high flowers —and superimposed them over the modest scene in front of me. As the sun slipped behind the far mountain, the sky darkened another hue: a hard green at the edge of gray. The hermit thrush trilled a melody and stopped at a high, treble pitch, leaving a musical void.

Over the last several years, I've observed people reviving landscapes through holistic planned grazing on farms and ranches in Vermont, the Great Plains, and Mexico that are verdant and full of birds and butterflies while conventionally managed land nearby languishes. Here I could bring in sheep to replace the lawn mower and manage brush. I could put in more plants to attract pollinators. In rainy Vermont where it never dries out, it's easy for land to be just okay. But enhanced ecology on every scale matters. As Geoff Lawton, the permaculture designer who famously created a food forest in the Jordanian desert, says, "All the world's problems can be solved in a garden."

———————

My work as an environmental writer has had me living in two worlds. In one I follow leading-edge thinkers as they bring ecological insights to farms and rangelands—places like rural Zimbabwe or the vast, cactus-y corner of west Texas. I've had the chance to witness land through the eyes of those who know it intimately. The nuances of these places have become alive for me, as if the soil and water were protagonists in a larger story.

In my other life I sit at my desk, writing. The Taconic Mountains beyond the window are but a backdrop—green, white, or brownish, depending on the season. It's not that I'm disengaged with my surroundings. I certainly enjoy the view and the steadfastness of the hills. I garden, with non-brag-worthy results, pick coneflowers for the table, and, when fruit beckons, brave spiky blackberry brambles. But for the most part, I have regarded the state of our environs as constant, even inevitable—as if it could be no other way. This is the land of Grandma Moses and picture postcards, after all.

Meanwhile I see stirring a global movement focused on ecological rehabilitation. Advocates and practitioners are taking a broader look at ecological vitality—and the ways nature strives to heal herself. The boundaries between approaches are dissolving. There is a new eclecticism, the sense of a larger, better-stocked toolbox to draw from.

While I say I've been living in two worlds, lately it's more like three, for there is also the global ecological present. While writing my previous books, I could carve out mental space for the belief that all will be okay; I could dwell in the realm of solutions. But our ecological crisis has become sufficiently grave that none of us can stave off awareness that the planet is in peril. Regardless of grim projections for the future, we are here now. Fear can be self-fulfilling. What if we make our hopes and aspirations foremost?

Katherine Ottmers, who has revived land near Big Bend, Texas, puts it like this: "We can be the beavers on the landscape." (Beavers being known to create wetlands and concentrate fertility.) She means we can *regenerate*, not just preserve and conserve. It's not only activists lit up by regeneration but also restaurateurs, fashion brands, and investors. These folk are not waiting for government to step in or for business interests to decide in whose interest it is to restore ecological health. Like the beaver, they are driven to enrich soil and enhance water movement of their own volition.

In recent years mega-storms, mega-fires, and out-of-season heat waves have seized the world's attention. But there is a growing awareness that nature—including that humblest of substances, the soil—is also our best ally. According to The Nature Conservancy, "Nature is the sleeping giant in solving climate change."[2]

Of course nature hasn't been sleeping, *we* have. Rather than seeking to understand how, say, the planet actually manages heat, we have looked to technology, neglecting the importance of intact ecosystems. It is a huge irony that billions are spent on climate predictions, geoengineering fool's errands, and surveys on people's climate beliefs, and so little on bolstering nature.

Climatologist Dr. Millán Millán, retired director of the Center for Environmental Studies of the Mediterranean (CEAM), says most of today's weather extremes are due not to greenhouse heating but to "local impacts going wild." According to Millán, ecological restoration in pivotal areas can reinstate water circulation and bring weather patterns back toward long-standing norms. He says strategically revegetating even a small expanse of land can make a difference in the surrounding region. If that's not an argument for healing land, I don't know what is.

What is climate change, anyway? The expression's omnipresence leads us to not explore its meaning. Generally, the phrase is shorthand for global warming from too much carbon dioxide in the atmosphere due to the burning of fossil fuels, assuming that there's a straight line from CO_2 emissions to increased temperatures, crazy weather, and sea level rise. And yet climate is too multifaceted, too complex a phenomenon for a single metric to do it justice. It is not just carbon but some very elemental functions that are out of whack. Climate change is best understood as the manifestations of disrupted carbon, water, nutrient, and energy cycles.

These cycles are key here. All are in continual flux and influenced by our actions. And so this reframing imparts the possibility of steering us toward equilibrium. Take carbon. There is too much in the atmosphere (and the ocean) and not enough in the soil and living organisms where it belongs. For this we can thank fossil fuels, internal combustion, and ten thousand years of agriculture, particularly the last several decades during which the pace of mining the soil has sped up rapidly. This carbon is sorely missed: In soil, carbon has multiple functions, including sustaining microbial life and holding on to water. Why not bring that carbon back?

This is where ecological repair comes in. By definition, regenerative practices bring these cycles toward balance. Space mirrors and other techno-fixes may make good copy, but the actual processes that regulate and moderate climate derive from living ecosystems. This book gives a glimpse of a growing cohort of scientists, mavericks, and young people who are shifting their focus to align with nature.

Across the globe, people are doing the essential work of restoring land and ensuring that the green-blue orb we sail on remains fit for habitation. They are out on grass-strewn rangelands, on smallholder farms in remote villages, in tropical forests marked by fast-growing trees, on parched land that was once productive and its people proud. They are cultivating beauty and helping communities grow their own food in a restorative way. For example, the Minnesota-based Main Street Project brings regenerative poultry to the Pine Ridge Reservation in South Dakota as well as to villages in Guatemala; their agroforestry model builds on the chicken's natural ecological niche as a jungle bird, enjoying protection and shade from the canopy.[3] We've been trained to believe that finding solutions is a job for the experts, but many unsung eco-restorers are doing the work, often drawing inspiration from clever creatures like ants, butterflies, and dung beetles.

Earth repair is a participatory sport: a grassroots response to evolving global crises. It is the inverse of apathy and an antidote to despair. Too often, activism is framed around being "against" something. This keeps us stuck in reaction mode instead of reaching creatively toward solutions.

Restoration can begin anywhere. Damaged ecosystems can be rejuvenated at all scales, from a small plot between sidewalk and curb to areas large enough to be labeled on a world map. We can all find a garden to tend. At the very least, none of us need sign on to the biological desert that is the typical American lawn. Imagine if all that land—the size equivalent of Arkansas—became carbon-rich, water-absorbing soils sustaining biodiversity above and below ground.

Regenerative work is creative and often surprising. Unlike the genius-in-the-garage narrative of technological breakthroughs, it entails an understanding of place and respect for community. This includes communities of too-small-to-see partners like microbes and mycelia, as well as communities of people that may have different opinions or be resistant to change. This book explores eco-restoration in various ecological and social landscapes, highlighting different scenarios and strategies depending on the local environment, wildlife, and history—often reflecting colonialism and the legacy of industry agriculture.

My book *Cows Save the Planet* is a soil's-eye view of the world. In *Water in Plain Sight*, I explore land, biodiversity, and climate from the perspective of water. This book is a global tour of eco-restoration, with particular attention to energy. I don't mean energy as in that which heats our homes or propels our cars. Rather, I'm referring to energy in its basic sense: what happens to solar radiation when it meets the ground. At its core, eco-restoration means improving a landscape (or seascape) to make the best use of solar energy to fuel plant and animal life. When we gauge the health of a landscape, the flow of solar energy is the game.

Picture arid African savanna that's been abandoned by farmers, an image that might accompany a news report on the ravages of drought. Still, day in day out the sun will shine. In this parched landscape, sunbeams fall on unwelcoming terrain: generating heat, evaporating moisture, killing whatever soil life persists. The sunlight remains dynamic energy, absorbed by the environment but its potential unrealized.

Imagine that this land is improved through holistic planned grazing, so that animal trampling, dunging, and urinating add nutrients and carbon to the soil, thus promoting native vegetation and maximizing water retention. The result? A portion of the sunlight is now captured by plants. The land is not yet fully healed, with grasses and thorn trees with canopies worth sitting under. But this sets the living system in motion as solar energy is (1) turned into sugar, some of which (2) is sent out through the roots, where it (3) supports a busy underground microbial community, and (4) is stored in the soil, as in a battery.

Now things get interesting, as even small ecological boosts spark further improvements such as more robust vegetation. This supports greater biodiversity, which, by keeping water on the land, leads to a restored water cycle—and a more productive landscape. Enhanced soil moisture means a longer period of green growth, and therefore higher production, a longer period of active photosynthesis and carbon drawdown, and a briefer season of dormancy and fire risk. Thriving landscapes enhance food security, which in turn means greater social stability; when people can grow their own food, they don't need to migrate to large cities where they may not be able to find a home or employment.

Peter Donovan of the Soil Carbon Coalition talks about "seven generations of sunlight," the fate of the vast amount of energy our neighboring star bestows upon the earth. He describes how this radiance: (1) is reflected back to space; (2) is absorbed as heat by the earth's surface; (3) evaporates water, driving the water cycle; (4) drives photosynthesis; (5) circulates through living things and drives biological processes. This fifth one he considers a kind of "knowing." It is what living organisms "do" with this energy. He notes that each subsequent generation represents less solar energy than the one prior.

The sixth generation is "consciousness, languages, beliefs," capacities we share with some mammals and perhaps some birds. Number seven is self-awareness, or, says Donovan, "consciousness about consciousness." This informs how we make decisions, how we engage with our own ignorance, assumptions, and beliefs.

This use of energy, he acknowledges, embodies the tiniest sliver of the sun's largesse. However, it paves the way for "creation, innovation, transcendence, and self-transformation." Says Donovan: "It generates enormous cumulative effects. Changes in how we see ourselves and our power relative to the biosphere as a whole, how we make decisions, recognize or fail to recognize threats and opportunities, and how we select and organize leadership are, and will continue to be, key influences on the power and uses of sunlight."[4]

We are at a point where human activity influences every ecosystem. In other words, human decisions inform the flow of sunlight. How do we embrace the possibilities inherent in that rarefied seventh generation—and apply creation, innovation, transcendence, and self-transformation to reversing the slide into environmental devastation? Can we facilitate the movement of energy so it creates abundance? Can these efforts bring disrupted carbon, water, nutrient, and energy cycles back toward balance?

Our land is an oddly shaped plot, its edges marked variously by barbed wire, road, and seasonal streams. It was once forest, likely a mix of pine, birch, cedar, oak, maple, hickory, and poplar ("popple" to locals), trees we find around here today. After Europeans arrived, much of the state was deforested to make room for agriculture, particularly raising sheep for wool. (In the mid-nineteenth century, there were six sheep for every Vermonter.) When the sheep industry tanked and moved west, where animals could be raised more cheaply, trees were lopped for the lumber trade and to get at materials like copper, granite, and marble. Our property was once a sheep farm, and we still have stone wall remnants.

If nature had her way, this land would revert to thick woods. But we're here, as are our neighbors and their houses, gardens, and pets. There are also animals that don't mind being around humans: deer, groundhogs, skunks, coyotes, raccoons, porcupines, and slinky, sly ones like fisher cats. We get a good bit of rain and snow, so it looks plenty fertile. Yet I know surface green can hide a multitude of ecological sins. I venture we could be doing better by our sunlight— and better serve the animals, birds, and insects that dwell here or migrate through.

We can definitely do better by our water. I learn this from Zachary Weiss of Elemental Ecosystems, who comes to assess how water flows here. He says anytime you have a house, you have impaired the local water cycle. One can "make amends" for one's water footprint, he says—the way someone might, say, offset the carbon dioxide produced by airplane travel: "You at least need to infiltrate all of the water that comes off the property to get to neutral."

This means finding a way to hold all the water that isn't seeping into the earth because your house, driveway, patio, and whatnot are in the way. He says landowners can add trees, create meanders, or devise weirs that retain water longer into the year on the property. And so: By our very presence we compromise how water runs on our mountain. What if water cycle impacts were a part of all building plans?

If our environmental challenges have taught us anything, it's that there is no "elsewhere." We are all connected in our foodsheds and watersheds. And we are certainly linked by the movement of wildlife, seeds, pollen, spores, and, alas, chemical toxins that pass through and over the terrain. While our ground is not degraded as some land-scapes are, why not take my own advice and start where I am?

In one fretful middle-of-the-night moment a few things seemed clear: that we are not going to make progress by arguing and blaming one another, or by sowing fear, or by pretending that everything is okay. The way we'll get there is with love: toward others and all living beings. For true change, people are mobilized by love more than fear. We humans have plenty of defenses for fear; our collective inept response to climate change has made this evident. Love emerges when the defenses fall away. I've come to believe that this is the heart of ecological restoration: bringing loving attention to a piece of the earth. We don't need anyone else's permission. We can start applying this at our own doorstep, wherever we are.

With this movement still young, there are few tidy stories and no formula. And so this book is also about process: how to grapple with the task of ecological repair and what mind-set drives it forward. We have no choice but to accept the unknown, as uncomfortable as it is. But discomfort can open us to new thinking, and new thinking can prompt innovation: exactly what's needed given that business as usual no longer serves us.

Sometimes it's just a matter of getting out of the way. Life forces will eventually kick in. But regeneration on a human timescale may require active engagement: using tools like the ax, the plow, and the cow. To quote conservationist Aldo Leopold, an ecosystem and its biodiversity may be restored by the "creative use of the same tools which have heretofore destroyed it."[5]

The book begins with rehabilitation on a large scale, in China's Loess Plateau. Next we see what three regenerative innovators are achieving under extreme conditions in the desert. In subsequent chapters—with visits to New Mexico, Norway, and Maui—we learn how ecological restoration is as much about people and indigenous knowledge as it is about tools and techniques. We then explore innovative efforts involving women, business, and youth activists.

We are called, too, to ask some challenging questions. Such as: How do we cross the threshold to a regenerative society? I pose this to permaculture teacher Andrew Millison, who runs the *Earth Repair Radio* podcast. His reply: "Some of the most successful eco-restoration projects have been in the harshest environments. We know what to do."Whether or not we get there, he says, "is about the human will and the human heart."

Which brings us back to our seventh generation of sunlight, the estimated 0.000001 watt per square meter that fuels self-awareness— in Peter Donovan's words, the "awareness of how we know, which can free us to shift our beliefs and behaviors."This is where we have tremendous leverage, he says, for "*creation is now* as well as in the past."[6]

The choice is clear, and it rests in our imagination and what we do with it. We can choose to imagine a blighted world, diminished of species and barely livable for humans (the planet we seem hell-bent on creating). Or we can imagine a world that is renewed in its verdant and abundance, the one the subjects of this book are working toward.

— ONE —

The Great Work of Our Time

Lessons from the Loess Plateau

Though the problems of the world are increasingly complex, the solutions remain embarrassingly simple.

— Bill Mollison (1928–2016)

Over more than four decades, John D. Liu has filmed in some ninety countries. As a network producer and cameraman, he has been witness to seminal episodes in contemporary history. He has filmed interviews with world leaders, people whose faces once dominated the news, were considered giants in their time, and have come and gone. He has crossed continents, walked through the corridors of power, and passed in and out of conflict zones. Yet for all that, the heart of John's lifework—his inspiration, his message, and his deepest hopes for the future of humanity and the natural world—can be distilled into four seconds of video.

This fleeting visual, quick as a finger-snap, depicts the restoration of China's Loess Plateau, a massive project that Liu documented over several years beginning in 1994. The time-lapsed sequence begins with a picture of terraces built into bare, sandy dirt with scarcely a scrap of green. The land structure is vast, even monumental, like an archaeological ruin. The people on the site are tiny figures standing and milling about on the flat surfaces.

One second into the film—don't blink!—plants begin to appear and it dawns on the viewer that there might actually be a landscape in there somewhere. Give it one more second and you've got trees: The ground has filled in and the setting looks more natural, less architectural. As second number four rolls by there is green everywhere, including in the distance, on terraced hills that had been bleak and dry.

A few short seconds on the screen and we've gone from wasteland to lush landscape. I've also seen this dramatic improvement presented as a single photo, in which from left to right the scene of barren hills visually melds into the verdant vista that it will become. In reality, this stunning makeover occurred over fourteen years and involved the hard work of tens of thousands of people, including those who appeared as flecks amid the edifice of lifeless dirt. But the transformation *did* happen, and that is the point.

———————

The Loess Plateau Watershed Rehabilitation Project was a huge undertaking, involving seven provinces, 2,137 villages, and a population of 1.24 million. It is a far cry from the grassroots, often improvisational restoration efforts I'll be featuring in the chapters that follow. And it is anything but a pitchfork-and-trowel backyard operation—though the terracing was initially done by hand, until small bulldozers were brought in that could do the job more quickly and cheaply.

I am beginning with this landmark project, however, because it shows the remarkable scale and speed at which ecological restoration can happen. It shows us that, given the opportunity, life will tend toward healing. Every community of living things—encompassing the largest trees and the tiniest microbes—strives to make the best use of sunlight in its particular locale. Given enough time, perhaps thousands or even millions of years, each will evolve to a highly developed and efficient system. The Loess Plateau project demonstrates how dramatically nature will rebound when we provide an additional nudge.

John Liu's encounter with the Loess Plateau project changed everything for him. Once you know that it is possible to rehabilitate large-scale damaged ecosystems, he says, how can it *not* be your life's purpose? How can you not do all that you can to spread this knowledge so that we can apply the tools of restoration to the open wounds of the world where land is degrading and people are suffering? To John, the understanding that ecological damage can be reversed obliges us to act upon that understanding.

John Liu is not shy about making such pronouncements. In the presentations he shares with international audiences, he stands before a large screen in his signature safari jacket and wide-brimmed canvas hat, images from his documentary work on ecological restoration scrolling as a backdrop. What the Loess Plateau project teaches us, he will say, is that "it is not necessary to accept degraded states." Since we know that large-scale rehabilitation is possible, "This knowledge is a responsibility. Human beings are required to understand this, because this is the determination of whether we can become sustainable and we can survive into future generations." He says these things, or close variants, in a calm, declarative voice touched by an affable midwestern tinge.

With the striking visuals, the rousing message, and his intrepid explorer persona—he has been dubbed "The Indiana Jones of Landscape Restoration"—John puts on a compelling show.[1] His statements unfurl with irrefutable logic: Once you've witnessed the visual narrative of the Loess Plateau's ecological redemption, our collective duty to restore natural systems is self-evident.

When I watch John speak, I often catch a hint of impatience beneath his wry, laid-back delivery. As in, *Where is everybody?* Soil is eroding at a clip of a soccer field's worth every five seconds.[2] Fires are choking cities from Los Angeles to Sydney, and causing devastation to wildlife. The current ecological trajectory is in no one's interest. Restoring ecosystems is, John declares, the great work of our time.

Certainly there are a lot of people around the world taking up the cause of earth repair. There is a global network of some ten thousand eco-villages that seeks ecological preservation and regeneration through empowering local communities. There are many long-running projects that offer models for reviving forests, including

Gaviotas in Colombia and Yacouba Sawadogo's work in Burkina Faso that inspired the film *The Man Who Stopped the Desert*. In hyper-arid Rajasthan, India, more than a thousand villages have replenished their water stores through simple water harvesting techniques, and farms now have as many as four crops a year where they used to have one. Some areas in Rajasthan have gone from 2 percent green to 48 percent green, and seen temperatures lowered 2°C (3.6°F).

All of this is positive, and merits acknowledgment.

But John Liu is thinking *everybody*.

Because despite many laudable efforts, the rate of ecological healing in the world is far outpaced by the rate of environmental destruction, and efforts to restore land have frequently been misguided. For example, many reforestation projects rely on monoculture tree plantations that fail to provide the ecological benefits of natural, diverse forests—or, in the case of eucalypt plantings, may cause problems by lowering the water table. Restoration programs require sensitivity to ecological nuance and the needs and aspirations of people living in the area. Ecosystem repair needs to happen in a timely way, but it also needs to be done right. The Loess Plateau project—and the examples to follow in this book—provides important lessons for how to go about this.

———

John Liu differs from many environmental scholars and communicators in that he emphasizes ecological *function* (how natural systems work) rather than *dysfunction* (when things go wrong); possibilities rather than impediments; processes rather than statistics. He talks about land degradation, biodiversity loss, and climate change as an expression of the state of humanity, as opposed to a misfortune that has somehow happened to us, or as the result of one simple cause such as unfettered use of fossil fuels. He has been known to say: "The state of our landscapes reflects the state of our consciousness." He sees the antidote to ecological crises within the human spirit. While many seek silver-bullet-type solutions, often through technology—as in the many headlines that ask, "Can This [fill in the blank] Help

Us Win the Fight Against Climate Change?"—John considers earth repair an opportunity for all. And he believes this is why ecosystem restoration offers so much promise to what ails human society as well as our beleaguered environment.

In short, John Liu regards the trajectory of ecological viability and human growth as a matter of intention. He says: If our intent is to wrest as much monetary wealth as possible from nature and turn this into products to be bought and sold—into "shiny objects"—then we will continue to degrade the environment and accentuate economic disparities. If, however, our intent is to support the ecological functions that yield us clean air, water, and food, then we will have a healthier environment as well as greater food security and human rights. The concept of intentionality is challenging to a public accustomed to outsourcing problem-solving to professionals. This framing can also be difficult for those (read: most of us in the Western world) reared on certain ways of understanding wealth and a set of assumptions about the march of human history. A look at John's personal and intellectual journey will reveal why this is—and why the concept of intentionality is so important.

All of which we find encapsulated in that brief pictorial sequence from China that Liu features in so many of his films. People engaged in agriculture were the reason that once-productive land was reduced to dried-out, denuded dirt. That said, people practicing agriculture were also what brought the land to its enhanced state, covered with vegetation and shimmering with life. This echoes Aldo Leopold's insight: people, livestock, farming tools, and land. The only difference was the intention.

Says John: "That ecological change means change for the worse is not a foregone conclusion. In our ongoing quest for food, shelter, connection, and meaning, we can act in a way that serves and even regenerates nature. It's up to us."

John Dennis Liu was born in Nashville, Tennessee, the son of a Chinese exile and, he says, a Southern Belle of Scottish descent. His father, John S. K. Liu, was from a landed family—not wealthy, but among the beneficiaries of several hundred years of feudalism. An officer in China's Nationalist Army, John Sr. took part in the Burma campaign, serving with the US and British joint forces. In 1943, while in command, he was hit by gunfire. As John tells it, his father stoically declared: "I'm dead. Leave me here!" He was in shock and wanted his men to flee to safer ground. The next in command was a noncommissioned officer, a strawberry farmer from Tennessee, who said, "You have just relinquished command. I'm in command. We're staying."

The NCO was able to get Liu safely out of the jungle. After evacuation to India and then Egypt, John Sr. was treated for his wounds in England. He then went to the United States, planning to make his way to California and return to China, a route that at the time was safer than traveling to China through Russia. During his journey, John Sr. stopped in Tennessee to see the soldier who saved his life, where he met Eleanor. They got married and had two children.

John describes his childhood in Bloomington, Indiana, as quiet and "rather special in a bucolic way." It was wooded and safe, and he could hike and camp in the forests and canoe on the rivers and streams. At thirteen his parents took John and his sister on an around-the-world trip by ship. At about this age he also began experimenting with a Super 8 movie camera. He went to college to study journalism, but his ambition got derailed when a professor told him that he would never work in the field. Discouraged, he dropped out. He played music, mostly piano, and drifted through his twenties.

In 1972, after Nixon's landmark trip to China reopened diplomatic relations, John Sr. said, "If he can go, I can go." He was determined to visit his mother, whom he hadn't seen in thirty years. With great effort he managed to secure travel documents and was among China's first international visitors in decades. When he returned to the States, he said to John, "You must go to China and help China develop."

John laughed at the idea.

John Sr. kept going back to China, and kept pressing John to go help the country join the modern world. Then in 1979, at a time when John was working with newly released video equipment, his father said, "You need to go to China now because your grandmother will die. If you don't go now, you will never know her." John considered this a better argument. He and his father traveled to China together for two months.

What John found was a society on the cusp of modernization, poised to emerge from poverty and isolation. "It was like going back in time. People didn't have cars. Everyone rode bicycles," he recalls. "They still had ration coupons for oil and grain and meat and cloth. There were lots of restrictions and trade embargoes. Sheep would be walking down the street in the middle of the night, with spray paint or a swatch of paint, saying they were headed to the Muslim quarter. To be food, you know, tomorrow's *shashlik* [lamb chops]." On top of the privation, there was widespread fear: "People didn't know who was going to inform on them. The Cultural Revolution was a horrible time for the Chinese. They don't even want to think about it now."

Once back in the States, John says, "I thought it would be much more interesting to film in China." He traveled back to study Chinese intensively. Upon his return he saw stacks of professional video equipment of the type he was using in the US: China had purchased the gear for 120 university studios, but no one knew how to use it. He ended up helping to create educational television studios in China at the Railway Institute in Changsha and then at the Beijing Language Institute. It was at the language institute that he met Kosima, a young East German student who became his wife.

At one point John's mother, Eleanor, sent him a newspaper clipping about US television networks opening offices in Beijing. "You should check this out," she wrote. Back then, even in cities the communications infrastructure was rough—a working phone line was a luxury—and Chinese-speaking journalists available to Western outlets were few. The field was wide open for someone like John: "I went to the CBS news bureau and said, 'Hi. I'm John Liu, and you need me.' And they said, 'You're hired.'"

John found himself in the heady, adrenaline-charged world of television journalism. He reported on the rapid developments in China:

"Rural reform, urban reform, the one-child policy and the student demonstrations, the presidents and the kings and the business leaders, and then of course the Tiananmen tragedy in 1989. For many years Americans and others would literally see China through my eyes. Then the collapse of the Soviet Union. I was sent to Moscow. My main attribute was that I just never quit. If the story was endlessly fascinating, I kept working day and night."

Over time John began to question the journalistic model. "Media didn't seem to want to take responsibility for the outcomes of their reporting and what was going on. They were always simply thinking that they were neutral observers and that's all they had to do and they weren't connected to the consequences," he says. "But being on the ground, I knew that if I went with a sixty-thousand-dollar camera into a devastated place, I've changed the situation. It is no longer the same place. People would act differently because I was there." The vaunted objectivity of journalism was an illusion, he came to believe, for reporters will inevitably approach situations from a perspective determined by their intellectual framework and cultural background. And then, of course, they get to pack up their notebooks and equipment and leave.

In journalism, too, John encountered the question of intention. Often, in his experience, what made an event a "story" was spectacle: the high drama that kept viewers glued to the newscast. "There's a difference between looking at the outcome of stories and the actual dust of the action and emotion of the moment," he says. To an editor, an "interesting" story in, say, a tribal area of Pakistan would likely involve some kind of violence or dispute. For John, interesting meant engaging with questions like, "Who are these people? How did they become who they are?"

Spending so much time overseas had changed him, he says, so that he had a sense of a global culture: the imperative to see ourselves as part of a connected, inclusive species. As for news reporting, he says, "We *need* to be interested in the outcome. We can't pretend we have no stake or that we're just observers. If you have no stake in the outcome, what are you doing here taking pictures of it and talking

about it? Is it just 'news porn'? Their suffering and war is not there for
the entertainment of people on the other side of the world."

———

The opportunity to film the Loess Plateau Watershed Rehabilitation
Project, beginning in 1995, gave John a story where the outcome
mattered: indeed, the well-being of millions of people depended on it.

This region in north-central China, named for its powdery,
windblown mineral-rich "loess" soil, was the birthplace of Chinese
agriculture, which emerged around ten thousand years ago, the same
time frame as the advent of settled farming in the Fertile Crescent.
As with the banks of the Nile, Tigris, and Euphrates Rivers, the
Loess Plateau—a swath of China the size of France—nurtured a
thriving civilization marked by advances in art, science, and philosophy. The region extends to several provinces, among the country's
most impoverished, and part of Inner Mongolia. It was home to
several dynasties, including the Qin, now known for the unearthed
life-sized Terracotta Army figures, and the Han, the most populous
ethnic group on the planet.

Loess soil forms over time from sedimentary material, transported
and deposited by the movement of glaciers. Although they're fertile,
loess soils erode quickly without organic matter and plant cover; the
Yellow River is "yellow" thanks to centuries of silty loess streaming into the riverbed. Over time, farming on hillsides, grazing, and
cutting down trees for firewood and land clearing accelerated the
erosion. At the height of the problem, each year one and a half billion
tons of soil and sediment were flowing into the Yellow River, causing
untold damage.

It is a narrative of decline we've seen repeated around the globe,
and the inescapable result was flooding, drought, mudslides, and
famine. People became desperately poor and, like all in tough
straits, did whatever they could to keep their livestock alive and
eke out one more harvest, even as removing more vegetation made
a bad situation worse. In John's films you can see weary farmers
push near-empty wagons of dried-out boughs across the dirt and

jaunty-horned white goats clambering down a barren slope. Every scene is brushed with dust.

In the early 1990s the Chinese government, the World Bank, and other NGO partners recognized a choice: let the degradation continue and consign tens of millions of people to perpetual deprivation, or address the underlying problem, which was the poor state of the land, and try to turn things around. And they chose to try to turn things around. It is hard to overstate just how herculean a task these entities were taking on. Nothing like this had ever been done before.

The first step of the Loess Plateau Watershed Rehabilitation Project was in-depth consultation with experts in fields from hydrology and forestry to agriculture and economics, and then, via China's Ministry of Water Resources, designing a plan, the basic tenets of which included an outright ban of ecologically destructive practices: deforestation, cultivation on steep hills, and free-range grazing. Sheep and goats that nibbled at will kept hitting the sparse vegetation until it could no longer grow back. In the immediate project area, local people were given land tenure rights and responsibilities and paid to do the work of creating terraces, putting in plants, and the like.

Juergen Voegele was the World Bank's study leader for the project and is now vice president for sustainable development. He says the project "started from a desperate situation in the Yellow River. This was an area that had experienced centuries of man-made degradation." It used to be forest, he says, but trees had been hacked down and burned ever since locals were defending themselves against Mongol invasions from the north. "In the 1980s and early 1990s, we would stand at the top of the hill looking down at totally bare valleys and say, 'What can we do?' It was bone dry and 90 percent of the vegetation was gone."

The good news, Voegele saw, was that despite centuries of erosion there was still topsoil left—fertile soil that wafts over from the Gobi Desert. "It took us about two or three years to actually understand what had happened to the land," he says. "It wasn't obvious. When you talked to people they would say, 'It's always been like this.' Even when you talk to an eighty-year-old farmer. 'Trees don't grow here. It's too dry.' We heard this from everyone. They were all completely wrong. It is not the case that local people always know better. They often forget. We didn't know, either."

The World Bank team came upon the crucible for rehabilitation by happenstance, Voegele recalls. They were scouting out success stories—clues to what might work. One day in a small village, a young woman showed them some walnut trees. The visiting group thought this was great: a benefit to people and the environment. Then they looked down into the ravine and saw it was bursting with green.

"There was grass and shrubs and trees," Voegele recalls. "Why would that be? Did they plant? Had they irrigated? What had happened was that when the woman had brought seedlings, the goats and sheep tried to eat the plants. They had held a village meeting and decided they couldn't have both. They then sold their sheep and goats. Everything else around you looks like the moon, and the only difference is that here there are not sheep and goats. Same soil, same temperature, same rainfall."

The quest to find an example of success led the team to a solution that might have eluded them had they focused solely on what went wrong. This reflects an approach that I have seen among other advocates of regenerative land management: Look at the positive outliers and learn from what they are doing. Researchers often hew to averages and dismiss any results outside the norm as flukes, a scenario that may discourage or obscure useful innovations. In China, Voegele's group analyzed the successful program to understand what held others back—and provide alternatives. In his words: "This isn't about what went wrong. This is about how to get this right."

Voegele and his team drew on what they learned to instruct others on how to improve their land. For example, they worked with other villages to better manage animals, introducing pen feeding so that livestock wouldn't free-graze on vulnerable terrain. They also brought in breeds of fat-tailed sheep that were more robust and productive and better adapted to the environment. They compensated people to keep their sheep and goats off the slopes.

After three years they started seeing startling results across large areas. Says Voegele: "Degraded land would re-green. Grasses would be three feet tall. We could do reforestation by keeping out animals."

Voegele takes several lessons from the project. "One of the most important things is that everybody had completely underestimated the ability of an ecosystem to restore itself. That is a phenomenal lesson."

In addition, he says, "The trick is not to know the answer before you know what the problem is."

The world's top geology, hydrology, and botany experts with the Loess Plateau project didn't have the answer, nor did the local people. Everyone assumed the main task was reforestation. Imagine how well that would have gone if the sheep and goats—vast herds, which Voegele says numbered in the millions—were still milling about and troubling the saplings.

With the animal challenge addressed, the project expanded in scope and complexity. Water harvesting (simple check dams to slow and capture water), reforestation, agroforestry, and revegetation measures were put into place, with an emphasis on perennial polycultures. In addition, land was specifically designated "ecological" (conserved) or "economic" (cultivated). As John notes, this last clarification "was based on an understanding that the ecological function was vastly more valuable than the production in the marginal lands." Trying to squeeze productivity from degraded land not only is bound to fail, but also means losing the environmental benefits, like improved water cycling and buffers against weather extremes, that land returned to nature will provide.

Of course, the fate of the land is inextricable from the fate of the people on the land. According to the World Bank, two and a half million people in the rehabilitated area were lifted out of poverty.

Liu Deng Fu is one of those people. He lives in Ho Jia Gou (Ho Family Gully) in Shānxi Province, less a village than the location of a long erosion ditch. In John Liu's film *Lessons of the Loess Plateau*, Liu Deng Fu describes how his family had come to the area in the early 1960s due to famine in the north. He says: "My mother didn't want to come to this place. When we arrived there was no food at all." Within a few years, his mother and little brother had died.[3] In the film we also meet Zhang Fang, who married into the Ho Family Gully and as a young bride moved with her new husband to a traditional cave dwelling built into the loess cliff. She says the area was bleak, with no trees or grass. "It was just barren hills and yellow like this soil here," she says, gesturing at the dirt around her that constitutes her home.

People here have found new sources of sustenance in recent years, however. They are growing fruit in orchards and vegetables in greenhouses; they are wheeling wagons laden with hefty cabbages and selling produce in a bustling market. We see attentive, well-nourished children sit in a bright, clean classroom. A man tells John, "Before, my dream was just to live in a brick house. Now it's as if our dreams couldn't keep up with the changes." No longer are people weighted down by the expectation, all too often realized, that their lives would only get worse.

Liu Deng Fu, who had until recently borne a life of hardship, is proud of the apple trees that have earned him more than ten thousand yuan over the past five years. He runs his fingers over an apple's shiny green skin and says his fruit is so good that local people choose it first. "It's hard to recognize this place now," he says. "We are now able to grow more crops and trees. It's really a big change. Before, we didn't have enough income for electricity. Now we have electricity and running water." The camera pans across children of various ages, eating and smiling and displaying good teeth. He says: "If my children's lives are good, then my life will be good, too."

Zhang Fang sits up tall and wears a crisp peach-colored jacket. Before the restoration she lived beneath the poverty line, earning between five and six hundred yuan a year. Now she earns between five and six thousand yuan. "After terracing the land, there is more water in the soil and we can grow many more vegetables like peas and corn." She has a surplus, which she sells or feeds to her pigs. She celebrates at her son's wedding. Family and friends blast horns, bang cymbals, and light firecrackers. She is happy to invite her new daughter-in-law, a pretty girl in a formfitting persimmon-hued dress and flowers in her hair, into her newly built home.

The World Bank has highlighted the rehabilitation project's many benefits: Farmers' incomes doubled, and the number of people living below the poverty line dropped by half; trees, shrubs, and grasslands regenerated naturally in some areas and were replanted in others; employment went up, with new opportunities for women; increased vegetation meant less erosion and sediment flowing into the Yellow River, as well as less regional flooding. In the words of one report, "Even in the lifetime of the project, the ecological balance was restored in a vast area considered by many to be beyond help."⁴

In setting aside conserved "ecological" land, the areas under cultivation achieved increased production. This is counterintuitive, since less land was devoted to farming, but the preserved areas allow nature to do its part. The gains resulting from ecosystem function may be indirect and may even be invisible. But they are significant.

Voegele had a close-up view of the change in people's living conditions. At the outset, he says, "Sorghum and millet yields were so low that it wasn't worth harvesting. Now this is one of the world's biggest apple-producing regions. When I went there the first time, people lived in caves. They would dig a hole in the mud and put a door outside. I remember stories from the villages of grandmothers who would stop eating and die so kids could eat." Thankfully, the scenario today is very different, he says, calling it "a massive transformation, socially and economically."

I was surprised to learn from Juergen Voegele that John's Loess Plateau films—which have not only been seen by millions of people, but have also influenced nationwide policies, including in Rwanda— happened by chance. Juergen Voegele and John had children in the same class at school (like John's wife, Kosima, he is from Germany). Through the children, Voegele learned that John was making environmental films and invited him to come to the area and take some footage. "There was not much to see, just some terracing," Voegele recalls. "We didn't think much of it, just some nice footage." The point, he says, is that "nobody believed that our efforts would make a dent. We wouldn't have bothered filming it."

As impacts of the project became apparent, the World Bank hired John to continue filming. When he returned to the same area in 2009, he had crew members that had worked on David Attenborough's films. In 1995, says Voegele, John was "sweeping the camera across the valley with not a tree in sight. Then, the place is green. That tells you the power of the story. Of course, the effort was human-supported. But it was nature that actually did it."

Matthias (Ties) Van der Hoeven is a young Dutch engineer who visited the Loess Plateau with John in 2016. Ties (pronounced *tees*) says that along the terraces there were "trees all around." At one point, he says, they visited a beautiful temple and he wandered off down below the terracing. John, startled, then said, "Ties, why are you walking there?"

Ties had replied: "I want to know if you're right." He told John he was looking for signs that the landscape was holding moisture, since John had said this would indicate a trajectory toward ecological health. "I said, 'John, I want to look down there because I saw there was a river flowing.' So we walked all the way down there. It was a bit of a nasty route. And we saw that the water was dripping." He said the soil was rich—"mulchy"—and covered in moss. This suggests that water was infiltrating and staying in place, which meant the soil would stay in place and support vegetation rather than blowing away or running off.

Ties has always loved the outdoors—as a child he hiked in the mountains and sailed all summer with his parents—but hadn't looked at the world from an ecological perspective. He worked for the dredging industry, a prominent sector in the Low Countries. He was starting to understand that dredging could serve ecosystem function as well as the needs of commerce. This is why he had come to John's attention, and when John said, "Ties, you need to come to China," he went.

Their last day in Yan'an saw a heavy rain, after which they went to the banks of the Yellow River. Ties noticed a pool of water bordered by grass and knelt down to take photos. "Local Chinese were standing around and looking at me, thinking, *What the heck is this guy doing looking at a pool of mud?* And then John saw me taking pictures of algae. He said, 'Ah! You understand!' And I said, 'Yes! I get it!'"

It wasn't just the algae, which generates oxygen and fertility, that had caught his attention. "It was the fact that I could see those skinny clouds at the top of the trees at the end of the day where you saw that water was recycling. It was the birds. It was the fact that nature came back so rapidly, and the scale of it. It was the water slowly dripping down that made me know that this was a ripe place. It was life-changing in that sense."

The tab for the entire Loess Plateau Rehabilitation Project was $500 million, furnished through World Bank loans and other financial instruments. If we break that down to 1,560,000 hectares and divide that sum by ten years, this amounts to about seventeen dollars US per hectare per year (less than seven dollars per acre). *Seventeen dollars US per hectare per year.* Given the impact, and the considerable costs avoided by decreasing erosion and flooding and less social welfare required, this is an extraordinary bargain. Or as permaculture designer Rhamis Kent, who works in degraded dryland areas in Southern Europe and North Africa, describes it (paraphrasing permaculture pioneer Bill Mollison), the solutions to our problems are "embarrassingly simple and embarrassingly inexpensive."

After witnessing the plateau's metamorphosis from wasteland to a steadily improving, productive environment, John's mission changed. While geopolitical events, the ups and downs of national fortunes, may seem important at a given time, he came to believe that they pale in significance compared with what happens to the earth. Indeed, he came to see many newsworthy incidents—flood, famine, contention over resources—as the result of environmental degradation. He considers this understanding crucial as people wake up to the challenge of human-induced climate change, which he regards as a culmination of the degradation of the world's ecosystems and, by extension, nature's capacity to regulate climate. He had seen that it was possible to restore ecological function to large-scale damaged areas to reverse this downward trajectory—and to do so within the space of a decade.

In order to tell the story, however, he wanted to better understand ecological processes. So John immersed himself in studying soil, hydrology, chemistry, and geology, particularly in healthy systems. He says, "Everyone studies dysfunction. It's a lot more enjoyable and rewarding to study function. Dysfunction is not inevitable, but you have to understand it." He found little academic research that looked at ecosystem restoration in various environments and set out to fill in the gap. He secured fellowships and published academic papers, many of which have been translated into other languages. Years as a television producer and cameraman, it turned out, proved to be good training in ecological observation.

He delved deeper into the Loess Plateau's story in light of the waning of other civilizations—the Maya, Mesopotamia, Rome—and applied a sort of "ecological forensics" to understand how the growth and apparent success of human societies seemed to lead inexorably to the demise of the natural environment that had enabled them to flourish. Environmental degradation, he says, is "connected to the behaviors of people. The Loess Plateau is an almost perfect example of the fact that human activity without ecological understanding leads to ecosystem collapse."

Through deforestation, overgrazing, and topsoil loss, he explains, "the hydrological cycle is destroyed. Instead of being nourishing, moisture becomes a kinetic force. And this kinetic energy is huge. When you carry that forward for decades and centuries and millennia, it has the effect of completely altering the landscape. What we're really seeing often is not a natural system. Rather, we're looking at a cultural outcome. When we call things like flooding 'natural disasters,' we may be completely wrong."

He thought a great deal about change: change over evolutionary time, those three-billion-plus years across geological eras as our big

rock in space became colonized by life, and change over historical time that has been observed and recorded by humankind. He
concluded that evolutionary trends lead toward greater complexity
and higher function; this is how the planet has come to support such
rich and varied ecological niches. Historical trends, when society acts
without ecological awareness, run in the opposite direction. On the
Loess Plateau, for example, human activity led to the landscape being
simplified: with scant vegetation, rain carving out channels, topsoil
streaming away, and life becoming less able to take hold. All of this
led to a cycle of diminishing function. In John's concise phrasing:
"Function is an evolutionary outcome; dysfunction is cultural."

The heart of a successful restoration effort requires allying with
evolutionary tendencies, the processes that lead to enhanced diversity,
intricacy, and redundancy. Upon studying and filming sites around
the world, John determined that the path to healthier, more resilient
ecosystems comes down to three interrelated trends:

1. Increased biomass
2. Increased soil organic matter [5]
3. Increased biodiversity

He writes that these factors are the basis for "processes that
create, constantly filter and continuously renew the atmosphere, the
hydrological cycle, and soil fertility as well as nurturing diverse life
forms."[6] When these factors trend up, the state of the environment
improves, creating the conditions for breathable air, drinkable water,
climate modulation, and food security. Reverse course and ecological,
economic, and, ultimately, social stability will deteriorate.

John's message, the enduring lesson of the Loess Plateau, is that
we have a choice. As for the consequences of this choice, he does
not mince words: "The outcome of human activity that doesn't align
with the evolutionary trends that determine ecological function leads
to the collapse of the earth's ecosystems and the end of civilization. Conversely, human activity that aligns with natural processes
can restore ecological function and thus sustain humanity. What

has impressed me is the difference between natural systems, which have huge organic layers, and human systems, which are massively degraded and actually have lost their organic material."

Which may remain the case—until we choose to align ourselves with processes that restore ecological function. Fortunately, life strives relentlessly in this direction, and will do so if given the chance. The rehabilitation in China provides us with a template. The specifics will vary according to the landscape, but the basic principles remain the same.

The initial task, says John, is to keep water in the landscape. Ensuring water infiltration "is the first thing, so that there's retention of moisture. This affects everything: evapotranspiration, air currents, the basic building blocks of function." The three foundational developments— more soil organic matter, more biomass, and more biodiversity—serve to slow the movement of water and keep it in place.[7]

Organic matter, which is mostly carbon, acts as a sponge; every gram of carbon in the soil represents eight grams of water that can be held. As for biomass, an incredible amount of water is embodied in plant and animal life, and this is moisture that continues to circulate in the local environment. An enhanced water cycle supports biodiversity, but biodiversity also supports a healthy water cycle. Organisms such as earthworms, dung beetles, and prairie dogs create pathways in the ground that encourage water to meander, while dam-constructing beavers create wetlands and slow water flow. These factors all contribute to the soil's water infrastructure, which ensures that water remains a nurturing element as opposed to a destructive force.

At the same time, modulating the flow of water increases soil organic matter, biomass, and biodiversity This is a virtuous cycle, with one positive outcome building upon the other. In the Loess Plateau Watershed Rehabilitation Project, people built terraces and small check dams to slow water runoff. Ideally, over time these structures will become redundant: The richer soil, ample biomass, and greater biodiversity will regulate water flow. Rather than course into gullies, water will wind through the landscape. Even tiny soil organisms play a role in creating pore spaces that keep water from speeding past before serving the plants and animals that depend on it.

John understood that more organic matter, biomass, and biodiversity don't just improve immediate conditions—but also mean advancing to a higher ecological state. This is the ecological concept of "succession," the progression over time toward greater complexity and resilience. One observes this biophysical narrative in the final two seconds of the four-second Loess Plateau clip, at which point the trees are established and the dominant hue is green rather than beige. We can see that the story of succession continued as Ties Van der Hoeven visited the site and observed the fertile soil, the bird life, the lofty clouds, and the near-spiritual sense of ripeness unfolding.

I came to writing about ecology by way of economics. Around the time of the 2008 financial crash, I was, like many people, underemployed with a lot of time on my hands. I started asking questions like, *What is wealth?* While driving my son to and from school, I would listen to news on the radio and experience waves of cognitive dissonance as announcers noted unemployment numbers as if reading a list of the dead. I wondered: *There must be other ways for people to be economically active. Why should everything depend on what we've decided to call "jobs"?* My questions led to freelance assignments in which I explored topics like, *Is the gross domestic product the best measure of prosperity? Why do local purchases help the economy?* and *Why are impacts to nature not figured into the bottom line?*

This last point represents such a breach with reality that it made my head hurt. It is due solely to the economic concept of "externalities." In a deft accounting trick, damaging consequences of conducting business, which might include water and air pollution, landscape degradation, and blighted downtowns, are simply not considered in a company's tally. The idea is that if it's not in the books, it doesn't exist. I saw this as fiction and sought to highlight the relationship between economics and the environment. The massive disconnect between the conventional economic wisdom of limitless growth and a finite planet is evidence of a serious design glitch: a society-wide exercise in fooling ourselves.

And yet this system, flaws and all, is the operating program for, well, the entire developed world. Its theme music—the vicissitudes of the market, jobs numbers, profits and losses—sets both melody and tone for newscasts and public debate, the inadvertent soundscape of contemporary life. It seemed to me essential to appreciate the difference between human-devised systems, which are negotiable, and the exigencies of nature, which are not. It has baffled me that many people regard it as the other way around. To wit: Author Raj Patel, who most recently co-wrote *A History of the World in Seven Cheap Things: A Guide to Capitalism, Nature, and the Future of the Planet*, said in an interview that many people find it easier to imagine the end of the world than the end of capitalism.[8] Reading that sent a chill through me: Crazy as this sounded it rang true, to the point of seeming obvious. Really, though? Is this the best we can do? Are we humans that lacking in imagination?

Questioning the viability of the program that runs the show—the commercial waters we are all swimming in—was exhilarating in some ways, but also disorienting. Aside from books (Jane Jacobs's *Cities and the Wealth of Nations* was a favorite), it was also kind of lonely. I sought out kindred thinkers and found myself drawn to those who articulated a critique that resonated, ones who called to bring economics in line with the natural world.

This was how I happened upon John Liu. He spoke about "naturalizing the economy" rather than financializing nature, which risks turning clean water and air into commodities. He made aphoristic statements like, "By valuing derivatives [products derived from extracting nature's bounty] more than the source of life, we have created a perverse incentive to destroy ecosystems." And, "Wealth derives from healthy ecosystems, not the production and consumptions of goods and services." His words had an oracular quality, unfolding with such fluency you could almost hum to his lines. He was saying pretty much what I wanted to say, but I hadn't been able to express it so, well, *economically*.

I got a chance to meet John, who is based in Beijing, at international events, including at COP21 in Paris. He was then filming interviews and staying outside the city in a chateau. A small evening get-together turned into a large gathering that filled the back room of a Chinese restaurant, where we were happy for John to order in Chinese for us. One thing he talked about over dinner was this idea that we could encourage people to go camping to restore the earth. "Glamping" was the term he used.

In fall 2016 John put out word on social media that he was giving a presentation in New York City, and was open to setting up visits and lectures in the region. I hosted him for two days in Vermont, where he gave a talk at Spirit Hollow, an ecological and spiritual learning center nearby. In true Vermont fashion we shared a potluck meal and sat around a circle in a yurt.

That evening proved memorable for another reason. Later on our dog, Tsotsi, came inside after tussling with a skunk and, after a moment of deceptive calm, overwhelmed the house with the most outrageous stink. John, Tony, and I rummaged through the netherworld of under-the-sink cabinets searching for home remedies, each of which proved futile against the reeking spray. That experience bonded us; John later sent an email from London to say that despite laundering his clothes multiple times, he thinks of Vermont whenever he opens his suitcase. It also served as an equalizer: Even your intellectual heroes are at the mercy of a stealth skunk attack.

My thinking about economics mostly emerged via reading and asking questions, some of which became the basis for articles and columns. John developed his more practically, while he was exploring ecological principles, from traveling the world and documenting the conditions of people's lives. One place in particular that honed his insights on wealth and economies was the Inner Niger Delta in Mali.

In 2010 John traveled to Mali as a fellow with staff from the International Union for the Conservation of Nature (IUCN) and Wetlands International. Beginning at the town of Mopti, the group traveled for two days in hand-wrought wooden boats, riding the wide ephemeral river that surges in the rainy season and recedes in the dry. From his telling, the journey sounds magical: They floated on the river by day and pitched up tents by the banks at night. They ate fish caught from the river and fruit gathered on stops. They passed fishermen casting their nets, and boats painted bright colors to lure paying passengers: a few pennies for the ride upriver. They saw long-necked waterbirds that may have flown halfway across the world in order to breed there.

Thanks to the bounty of the monsoon rains from the Guinea Highlands, the delta floods every year. This turns the usually narrow and staid Niger River into a sprawling water corridor that transforms the arid landscape into wetlands and a means of travel. The Mopti harbor had been busy as groups of people matter-of-factly piled all their belongings on the boats and set off, the river carrying them away toward Timbuktu. At the height of the inundation, around September, as much as thirty thousand square kilometers (11,600 square miles) of land—an area larger than the state of Vermont—may be covered by water some six meters (twenty feet) deep.

The annual influx creates a temporary oasis, and life rushes in to meet it; the ebb and flow creates a rhythm that communities build their lives around. More than a million people, including fishermen, pastoralists, and farmers, are directly dependent on the annual flooding for their livelihoods. The wildlife—Nile crocodile, manatees and hippos, migratory birds, and fish, some seen nowhere else—is abundant. Among the plants, there are freshwater mangroves and "flood forests" whose trees have adapted to survive both in water and on dry land.

The vast, inundated floodplain and the sight of entire village populations carried by the swells left John stunned: "If I hadn't seen it, I don't think I would have believed what I saw." He says the delta's seasonal flows are significant to regional moisture regulation, and therefore climate. The current vulnerability of this powerful system alarmed him. Upriver dams and irrigation projects and a decrease in rain are now altering once-dependable patterns; indeed, the desert landscape of the Sahel is creeping down from the north. The region has seen violence from clashes with extremist groups and Tuareg rebels. Already in turmoil when John was there, since that time the area is often deemed too dangerous for scientists and other visitors.

To John, Mali represented a paradox. This is a country of eighteen million people in an area nearly twice the size of France. It is blessed with natural wealth yet is among the poorest nations in the world, with associated abysmal life expectancies and infant mortality rates. "The evolutionary nature of this place is unique and its cycling of water contributes to climate regulation, but we don't value it," he says. "How can we consign to poverty those whose responsibility is to maintain these natural systems?" The fact that the region's people are relegated to poverty is, he believes, the crux of the matter: Precisely because the Inner Niger Delta ecosystem is not valued, the local population is given little choice but to degrade it.

"The message we give to people in less developed countries is that they have to have growth, which means energy use, factories, industry. We're not thinking of the ecology of the system," he said. "People don't know what to do. In order to participate in the global economy, they've got to have something to sell. Which often means they've got to destroy something. If the product is, say, frankincense, then they may be pulling up the last of the plants that are barely holding the sand in place. Because no value is given to what they have, they're supposed to want what we have in America or Europe."

In Mali the way our economic model mismeasures wealth became clear to him: "To insist that fragile but important ecological zones have no value unless they produce goods and services bought and sold by the global economy is simply wrong. I saw that the value of ecological function was vastly more than the products of the ecology. And by valuing the derivatives higher than the ecology, you create poverty. It hadn't occurred to me before that these people had anything of value unless they pulled it out of the ground and fashioned it into something. But I went out there and thought, *Oh my goodness, I'm wrong.*"

Applying a conventional development perspective to land-rich, cash-poor countries can perpetuate a scenario in which ecological degradation and human impoverishment build on each other. One of the research team's tasks in Mali was to assess the environmental impact of dams that regulate water for agriculture and hydroelectric power and to reduce flooding. Big irrigation and power projects appeal to governments and businesses within the region and abroad (often via land grabs) but can have devastating consequences to those who depend on that water. As Wetlands International says, with diplomatic understatement, "All the dams studied provide benefits to specific groups of people and bring costs for others."[9]

The researchers got a vivid look at how dams affect communities downstream. "The people there are stressed because they were trying to survive on fishing but lots of things were starting to disrupt the water," John says. "When they stopped having the fish, they started going after birds. So these rare and precious birds that fly down from Siberia each year are being bought and sold like chickens. The people used to be fishermen and were used to netting things. They would have nets out hanging on bamboo poles. So now they flush the birds out of the bushes and get them to fly in the nets. Poor women were selling food at the lunch counter in the village, *selling rare cranes*."

For John, this epitomized the folly of promoting enterprise to help people in struggling communities without an understanding of the ecological implications. Activity that can undermine ecological function, such as diverting water for upstream farms or impromptu waterfowl barbecues, often has consequences well beyond the local area. The Inner Niger Delta is central to conveying moisture throughout the region. This in itself has huge significance in West Africa. But there is much still to be learned about the delta's global role as a hydrological pump. The more we understand about how water moves through vegetation and through the atmosphere—such as the "flying river" above the Amazon rainforest that carries more moisture than the Amazon River itself—the more we discover that its significance transcends what we've been able to quantify.

This water system, John says, "is being degraded because no one bothered to value what it was. I think if it was valued and it was cared for, we would get a completely different result. We're now at the point where we're going to lose that. They are eating the rare birds!" He believes that if as a society we recognized other ways to create wealth than to produce and sell things—if we saw wealth inherent in thriving ecosystems—there would be a greater impetus to protect and restore the land- and waterscape. And this would benefit us all.

Mali's example shows how market economics often puts people in cash-strapped countries in the untenable position of needing to destroy natural wealth in order to survive. However, placing value on ecosystem function would open up the possibility of environmental restoration as employment. John points out that across the developing world, in degrading rural areas and the frayed margins of megacities in the Global South, huge numbers of youth languish without hope of viable jobs, a scenario that contributes to global strife and disaffection. Depending on the area, ameliorating this could involve a variety of efforts to establish conditions for more sustainable food production, like the terracing and check dams used on the Loess Plateau. Indeed, what more rewarding livelihood for people alienated from capital markets than to restore their own landscapes and lift their communities out of destitution and despair?

To get there we need a shift in the way we think about work and productivity. Sounds like a tall order. Yet today, beneath the background murmurings of stock averages and the fates of iconic brands, people are increasingly questioning the assumptions that underlie common metrics. And many are exploring ways to reconcile human and ecological needs. Says John: "Areas now seen as 'poor' and 'underdeveloped' are actually the areas of most potential because restoration could engage people in meaningful work and make a huge difference."

I often wonder about the way our society talks about jobs, and believe we've got things backward. Rather than asking what work needs to be done or would make life more pleasing and fulfilling, we begin with the question, *What can a person do to make money?* The answers we come up with justify all sorts of behavior that ultimately harm people, communities, or the environment. As a result we've got people doing work that does not better the world or provide any benefit beyond yielding a paycheck—and may be linked to ecologically destructive extractive or polluting enterprises. And we see this as okay because this kind of employment fits into our framework of what constitutes "a job." When it comes to the activities of industries of an environmentally dubious nature, how many times do we hear the refrain, *but we need the jobs*?

This reactive approach deprives many people of satisfying work. While everyone wants to contribute to society, many suspect that their eight-hours-a-day pursuit is pointless. As David Graeber, a professor of anthropology at the London School of Economics and author of *Debt: The First 5,000 Years*, writes, "Huge swathes of people, in Europe and North America in particular, spend their entire working lives performing tasks they secretly believe do not really need to be performed," adding that "the moral and spiritual damage" of these "bullshit jobs" is enormous.[10] The prospect of viable careers in eco-restoration turns so many assumptions upside down. It offers the possibility of healing the human spirit as well as healing land. Why not promote work that benefits more than the employer and investors who profit from it? At the end of the day, we're all in this together.

"There is nothing wrong with the earth," John likes to say. As we know, earth's systems are far from static. They are dynamic and they continually adapt to changing impacts. The logic of nature is no secret; it is laid bare in every streambed, every handful of living soil, every spiderweb, if we bother to take a look. Its tale is told through accrual or retrenchment of biomass, biodiversity, and soil organic matter: the stuff of life. We can acknowledge this and stride forward arm in arm with nature's inclinations, for we depend on it for well-being and survival.

John told me of his trip to Rwanda in 2006. One day the team was taken to see a memorial to the Rwandan genocide. On the way back it was hot so the group's jeeps had the windows open despite the dusty drive. John heard sounds that seemed to come from a stream within the mountain rainforest to one side of the road. He stopped the convoy, grabbed his small camera, and pushed his way into the jungle.

There was no path, so this entailed climbing over sticks and vines and pulling off sticky burrs while trying to keep his balance. Within ten minutes, maybe fifty meters (150 feet) from the pavement, he came to a waterfall. The spot was pristine. Every surface was covered with moss. There was a sweet, intoxicating smell, like very ripe fruit, and orchids dangled everywhere. As he glanced about he saw hundreds, even thousands of butterflies in all different colors. And this in a place that had endured a horrendous genocide but twelve years prior.

He walked back to where his young colleagues, unaware of the patch of paradise he'd just stumbled into, were waiting, smoking cigarettes. It was at that moment of contrasting realities when he was hit with the realization, one that serves as a kind of mantra for the task he has embraced: *There is nothing wrong with the earth.*

That experience under nature's spell highlighted something else: While traveling the road had been hot, the jungle clearing was refreshingly cool. John often emphasizes that ecosystem function moderates temperature: Plants cool their surroundings through transpiration, whereby plants release moisture in the form of vapor—as well by shade. Transpiration, the upward movement of water through plants, consumes heat and is therefore a cooling mechanism. Without plant cover, the soil (or asphalt) gets a direct hit of solar radiation, which is then absorbed. While it seems obvious, this aspect of healthy landscapes scarcely makes a dent in discussions of climate and heat waves.[11]

"Degraded lands exhibit massive temperature differential compared to functioning ecosystems," he says. Regarding climate, "if we physically restore vast areas of the planet, we are not just reducing carbon by bringing it into the soil, but also minimizing temperature differential." In a recent webinar, he said that measurements taken at China's Great Green Wall afforestation project showed a 45°C (81°F) variance between sand (70°C/158°F) and the treed shelter belt (25°C/77°F); in southern Africa the temperature difference between vegetated and degraded land was 20°C (36°F). As we explore climate solutions, we need to consider the cooling effect of plants and how they drive moisture.

The astounding fecundity of that jungle evoked another truth about nature: its proclivity to abundance. John refers to the fact that market economies are based on a perceived notion of scarceness (we're afraid we won't get what we need) or insufficiency (our clothes aren't sparkling clean so we need the new-and-improved detergent). Scarcity drives prices upward. Supply versus demand is the principle that keeps us buying and gets us paying more. "We've been saying that scarcity is the basis of wealth," he says. But if we take our cue from functioning ecosystems, "it's abundance that's the basis of wealth."

The diminishment of natural wealth that we're seeing now may be the predictable result of human behavior driven by greed, but we don't have to accept that. The outcome, he reminds us, turns on intention: "If our intention is to make the earth beautiful, that is what will happen." That beauty can return to such depleted terrain as the Loess Plateau is a testament to nature's capacity to heal—and the power of intention. This has implications for many parts of the globe where people are suffering and there is little life to be seen.

Environmental rehabilitation is a goal for which every one of us can play a part, he says, and poses the question: How might a shift in priorities toward environmental healing promote human healing—on an individual as well as a societal level?

— TWO —

Life Begets Life

Replenishing Middle Eastern Deserts

Every living thing is in a state of worship.
—the Qur'an

T he Al Baydha Project is a community development venture
fifty kilometers (thirty miles) south of Mecca that launched
in 2010. But according to Neal Spackman, the permaculture
designer who planned and managed the demonstration site, its story
really begins in 1952. This is when the trend toward Bedouin settlement began. These nomadic pastoralists, whose way of life revolved
around moving with their herds between desert and farmed areas,
were now increasingly tethered to one place, their portable goat-hair
tents giving way to cinder-block homes.

Neal, who has spent most of the last several years living and working in western Saudi Arabia, gives me a brief history lesson. "In 1952
there was a national Saudi policy that eliminated all of the traditional
tribal boundaries and said that all land belonged to the kingdom," he
says. Therefore any land not privately held fell under the purview of
one of the national ministries. Since Bedouin culture is not structured around private ownership, this meant that "the traditional form
of land management, called the *Himma*, was essentially abolished.
That system had maintained the fertility of tribal lands for a couple
of thousand years. It predates Islam. With Himma there were
rules about how many animals and whose animals were allowed on
which land, and how much land. Land was allocated for the poorest

members of the tribe and their flock." With the exception of those living in Mecca or Medina, the vast majority of people in this region had been nomadic pastoralists, he says. As of 1952 it became harder for the Bedouin to build their lives around their animals.

The result, says Neal, offers a cautionary tale about the unintended social and environmental consequences of policy: "The way that law was interpreted was that anybody was allowed to graze anything at any time." And so they do; or rather, so do those with the money to take advantage of prime fodder. When the area receives rain, wealthy residents of Mecca, Medina, and Taif—cities within a two-hour driving radius—truck in their camels, sheep, and goats to jump at the flush of grass.

"In Al Baydha the average person has twenty-five to thirty head of sheep, maybe ten camels or goats. The people who are trucking their herds around would have 150 camels and 600 sheep. So they come in, the herds eat up all the grass, and then they leave. And it doesn't rain for another year. What it means is that people who actually live in Al Baydha have to buy feed for their animals. The way they get cash to do that is by cutting down trees. They cut down trees and they sell it to be used as firewood or charcoal in restaurants in Mecca. Then they buy the imported hay from Australia and that's how they keep their animals alive and that's how they keep their culture alive." With tree cover mostly gone, the local people now depend on social welfare. Hence the project's emphasis on poverty reduction and housing.

The change in land access created something of a forage free-for-all. Which put the Bedouin in a situation where the only way to preserve tradition and sustain their herds was to further degrade their landscape. It is a predicament similar to the one John Liu witnessed among the people of rural Mali: The need to sell something for cash left them no choice but to exploit and therefore undermine local ecology. As Neal laments, "It's not an uncommon pattern."

It is a pattern that the Al Baydha project is designed to avert.

Saudi Arabia is known for its harsh, even punishing, landscape, yet older people's experiences suggest a more verdant past. "There's no data on this, but when I talk to people who are sixty and over, my sense is that almost everybody had a childhood place to visit that was very green and had water flowing most of the year—and now is just desolate," Neal says. "I specifically ask people: 'When you were a child, was there a place close to nature that you used to visit with your family that was lush, and that isn't anymore?' Nine out of ten times the answer is yes."

In short, this desert land, where years can elapse between rains and the summers are so hot one can work outside only at night, is more desert than it used to be. "Anecdotally it used to rain twice a year," Neal says. "Animals would eat grass in the winter and feed from trees in the summer. Shepherds would hit the leaves off trees in the summer for the animals to eat. There aren't trees anymore. They're gone, almost all of them."

For dry and "brittle" landscapes, the capacity to flourish turns on small margins. (*Brittleness* refers to the length of dry periods; Vermont, with no dry season, is a non-brittle environment.) The brittleness scale was articulated by Allan Savory, the Zimbabwean wildlife biologist known for developing Holistic Management, a decision-making framework applied to grazing systems. Savory observed that a landscape's response to management is influenced by the distribution of moisture throughout the year—a more complex metric than simply measuring total rainfall over time.

Particularly in highly brittle and hyper-arid terrain, where evaporation can exceed combined precipitation and condensation several times over, every drop of water matters. These conditions create specialized biological niches; in brittle ecosystems, living organisms are quite inventive about adapting to temperature and maximizing available moisture. In western Saudi Arabia, for example, where summer daytime temperatures can top 50°C (122°F), some native plants flower at night and rely on pollination from bats. The Namib beetle sips on droplets of fog by lifting its hindquarters so that the damp sea air condenses on its back and beads of water roll into its mouth. A slight shift in the delicate choreography that sustains these ecologies—the dynamics that keep the bats servicing night-blooming flowers and darkling beetles basking in fog—throws the entire system off balance. And so it begins to deteriorate.

That systemic decline is the hallmark of *desertification*: land's diminishing ability to support life. It is wicked easy to set a landscape on this path. Some of the more popular routes are (1) farming on hillsides, which leads to erosion, drought, and flooding, followed by more topsoil loss and persistent gullies; (2) overgrazing or, as frequent if counterintuitive, *under*grazing, for absent the biological impact of herbivores, including their cycling of nutrients, grasslands will degrade; and (3) cutting down trees.

According to the United Nations Convention to Combat Desertification (UNCCD), about one-third of the earth's land surface is either desertified or headed that way. (The United States is about average in its percentage.[1]) The downward slide can occur over a long time horizon, as was the case with the Loess Plateau, which took several thousand years to test the limits of habitability. Or rapidly, as when a few generations of overzealous cropping led to the US Dust Bowl.

History reveals that most deserts, including the Sahara and even the Atacama in South America, deemed the driest place on earth, were once vegetated. Which makes sense: Many oil-producing regions, including Saudi Arabia, are desert, and petroleum is a product of photosynthesis from the ancient past. This has raised the question of whether we have any "true," or original, deserts, or if these sere environments always point to the disruptive hand of humans. Factors independent of human behavior—including rain shadows, air circulation patterns, and ocean currents—can also contribute to the formation of deserts, but still, when unchecked, humanity's penchant to extract value from the earth leads to depletion and desertification. As evolution biologist and futurist Elisabet Sahtouris writes, "Biologically, we are a desert-making species."

But even such a severe assessment does not mean that this course—what Sahtouris calls "one-way entropy"—is inevitable. Sahtouris herself describes a seismic shift in worldview, in which a growing number of people see the planet as a living system—a departure from the mechanistic model that has dominated Western thinking for the past several hundred years. In her words, an understanding of life as continual self-creation highlights an awareness of nature as a "conscious collaborative process."[2]

This mind-set of collaborating with nature is core to the project of ecological restoration. It is this shift in perspective that enables one to see the potential for bountiful apple trees within the wasteland of the Loess Plateau; to imagine butterflies in the shadow of Rwanda's killing fields. What can one envision for the desert? What seeds of possibility lie dormant within this arid ground?

Here the drama of survival is seen in sharp relief; a desert represents a stark canvas upon which to portray an account of life as an emergent, self-realizing creation.

It is worth asking the question, *What is a desert?* Sure, we assume we know. My mind flits to cartoon images: Bugs Bunny, ears drooping in the heat, desperately crawling across endless sands to a rumored water source, marked by a lilting palm tree. The accepted definition refers to a lack of rainfall: Any place that receives less than twenty-five centimeters (ten inches) of rain per year is considered a desert.[3]

During an interview with Andrew Millison on *Earth Repair Radio*, Neal caught my attention when he suggested a novel way to characterize a desert: "A desert is a place where when it rains it floods." The rationale is that once the land is capable of absorbing the rain, it ceases to be a desert. It will then support life, beginning with the humblest forms: life microbes and fungi and insects that in turn create the conditions for more life. The wisdom here is that the amount of rain the land receives doesn't tell us how the land *works*. Neal's reply acknowledges the dynamism of biomes: that land is not static, but moving toward states of greater or lesser function. Land is always in the process of self-invention—or rather, co-creation.

According to Neal, the Al Baydha Project is generating water in the desert. Over the course of five years, his team planted forty-five hundred trees. By the middle of 2016 the project ran out of funds and the irrigation was cut off. Two years and a scant five centimeters (half an inch) of rain later, 80 percent of those trees are still alive. The irrigation water, harvested from rainfall via simple earthworks, has neither run off nor evaporated in the heat. Instead it has remained in the environment. Neal says at this point the trees are likely securing the water through mycelium networks at the roots. Let's assume this continues, with more trees and other plants circulating moisture. Will Al Baydha still be a desert?

––––––––––

Neal Spackman is an unlikely candidate for the task of restoring deserts. Now in his mid-thirties, he grew up in suburban Minnesota and then attended Brigham Young University, where he studied Arabic and the Middle East. His course of study was inspired by a Tunisian goldsmith he befriended while serving as a Latter Day Saints missionary in Guatemala City. The two had spoken in depth about Islam and Christianity, and this intrigued Neal; he'd had little engagement with Muslims. He was still abroad when 9/11 struck. In his quest to understand the complexities of what took place, he enrolled in classes in Arabic and Middle Eastern politics.

To become proficient in Arabic, Neal lived in a foreign-language residence at Brigham Young and then spent six months in Alexandria, Egypt. He recalls: "I had just gotten married so my wife and I did our honeymoon in Egypt. We lived in a high-rise next to a street market. Every day I'd go to school and then for five or six hours I'd play chess against a butcher or a guy who sold vegetables. We played chess and they would drink tea and help me read the newspaper. I absorbed as much Arabic as I could." His Egyptian friends looked out for his wife, Candice. They helped her cross the street through chaotic traffic and stood by her side at the market should anyone untrustworthy approach her. He says that as their Arabic improved and he and Candice became part of the community, the prices at the market suddenly dropped. They were no longer seen as tourists.

After college Neal worked for a firm in northern Virginia that provided news and cultural analyses of Arabic media. Which was, he says, "interesting but also boring." He pauses. "Maybe 10 percent was interesting and 90 percent boring." He quickly realized he did not want to work in a cubicle for the rest of his life. Meanwhile, he had been reading about permaculture and natural building and had dreams of starting a natural building company. Instead, he says, "The Saudi discussion came up."

Neal had become friendly with an Egyptian American neighbor, Mona Hamdy, long active in the nonprofit development world. She was then exploring a poverty reduction project with local Bedouin with Saudi princess Haifa bint Faisal, and submitted Neal's name for consideration. Through the interview process, Neal introduced the group to Geoff Lawton's Greening the Desert Project. The project features the transformation of a hyper-arid, highly saline four-hectare (ten-acre) plot of lifeless sand in Jordan's Dead Sea Valley into a productive food forest. A brief video of the venture—of achieving the impossible by bringing green to an inhospitable patch of desert with earthworks and careful planning—has inspired many people.[4] Neal raised the possibility of including a permaculture component in the Saudi development program.

The funders invited him to join the project. While Neal knew a lot about permaculture, he had yet to actually implement a design, so he requested their financial assistance to take a Permaculture Design Course (PDC) with Lawton. Late summer 2010 Neal quit his job, and two weeks later he moved to the Middle East. He says, "I don't think I was their first choice, but there wasn't anyone else who would actually go out and live with the Bedouin." Taking on the assignment and leaving home was not an easy decision; by this point he and Candice had two children. Neal says her encouragement was pivotal. "She said the right things. She said, 'This is the kind of thing you've wanted to do and the kind of thing you've dreamed about. I think you've got to do this.' I think she was very reluctant and very nervous."

Once Neal completed his PDC with Lawton in Jordan and the two were together in Saudi Arabia, they selected a forty-hectare (hundred-acre) demonstration site in Al Baydha and secured approval from tribal leaders and the local magistrate. After this, Neal says, "It was just me and Princess Haifa and the Bedouins." Even dealing with basic logistics required constant improvising. Sunday through Thursday he slept in a room in the magistrate's office close to the site. On the weekends he went to Jeddah, on the coast, and stayed at the palace where he could get online and be in touch with his family.

Neal recruited a team of fifteen local men. Once the degree of work involved became clear, the number quickly dropped to four: two older and two younger. "I think they stuck around mostly for the promise of a paycheck, at least initially," he says. It took a long time to build trust; he was the first non-Muslim and foreigner they had met. As for applying permaculture design, "The process was to look at all the natural resources available and the heritage of the people there. And then ask: What can we actually do? There was absolutely no biological capital to speak of. There were grazing animals, but no carrying capacity. We've got mountains, the foothills of the mountains that [the city of] Taif is built on. And so we have mountains and sand and rock. And camels and sheep and goats, but no grass. And we have floods. When it rains there, water collects in the mountains and runs down. So the general strategy was to slow down the water and the floods and get it into the ground instead of washing out, evaporating, or running into the Red Sea."

With "get the water in the ground" as the guiding principle, the team got to work. They devised a flood management and water harvesting system based mostly on stone earthworks. These took two years to build. "The vast majority were check dams and small stone barriers that were maybe ten to fifteen centimeters [four to six inches] tall, and that we raked onto contour," he says. By "contour" he means following the land's natural slope and working horizontally, rather than up-and-down, to slow the water so as to minimize runoff and erosion. In January 2011, after three months of solid work, rains strong enough to cause flooding in Jeddah gave them a chance to test the system. It would be three years before they had any meaningful rain again.

The heavy rains "washed out some of the early earthworks that we built and completely flooded our swales," Neal says. "We caught maybe seven thousand cubic meters of water, which is about seven million liters [close to two million gallons]. This led us to ask: What if we took seven million liters and amortized that over a three-year period? How much water can we use per day without draining what we just caught? In other words, we had a deposit of water into our bank account. How much could we withdraw on a daily basis without getting to zero?"

Establishing a water budget set parameters for what they could do without falling into water debt. "We plant trees and we irrigate them, but we're not going to use more water than what we've caught off of the floods," says Neal. "That is how we know we're not draining aquifers and groundwater. Based on that, maybe we can plant a thousand to fifteen hundred trees."

Depleting aquifers is not a small thing in Saudi Arabia, where several decades of feverish farming—despite its aridity and lack of surface water, the country was a wheat exporter for a period during the 1980s and '90s—have exhausted the nation's underground water stores.[5] The country has used up 80 percent of its groundwater in only the last fifty years.[6] "We tend to look at groundwater as a resource," he says. "Groundwater isn't a resource. Groundwater is just a storage facility." In other words, by lumping groundwater in with the rest of the water budget—by regarding it as fair game—you risk running down your water reserves. Like soil erosion, the mismanagement of groundwater is considered a quiet global threat.[7]

Over time the group settled into a routine, devoting most of their time to building earthworks. When it finally rained, they stopped and shifted to planting. Says Neal: "We'd measure the water we caught to estimate how much we could irrigate with that for the next three years, and then plant accordingly. Then we set up a drip irrigation system where we were measuring our daily rate of consumption and making sure we stayed within our water budget. Up to when we stopped irrigating in 2016, we had fifty thousand cubic meters of water that we captured from flooding. And we had used twenty thousand for irrigation. So we were putting twice as much water into the soil as we were taking out."

Among the species they planted were legume trees, which are useful for firewood, forage, and fixing nitrogen. These include mesquite, leucaena, and several types of acacia, in particular *Acacia tortilis*, or umbrella thorn, which is resilient in hot, dry, sandy conditions; fruit trees, including date palm, mulberry, and *ziziphus*, or jujube; and grasses, vines, herbs, ground covers, and plants that produce cash crops, such as moringa and frankincense. Neal specifies that each variety "had an economic component that we thought we could harvest, but really we didn't know what we'd be able to grow. We knew that initially the dominant ecology there was acacia-based. Two kinds, neither of which you would consider a strongly economic tree."

He devised the list of plants by looking at "climate analogues": identifying regions that have a similar climate, often in different parts of the world, and choosing species that do well in those places. For example, people in the California permaculture community are exploring plants from areas in Central Asia with comparable temperature and rainfall patterns. Farmers across the globe are now using climate analogues to identify plants that will thrive as the climate changes. The team was able to source all the plants they wanted from local nurseries and gardens.

With funding of the ecological development stalled, the Al Baydha site today consists of a project office and some other structures that are protected by a security guard. Suspending irrigation was certainly not in the original plan. But it has provided a real-time test. "The goal always was to recreate an ecology," Neal says. "This is kind of the permaculture ideal: that the ecology runs on its own and you just go in and harvest. I would guess 20 percent of our trees have died. This has given us a very clear picture of what we need to plant, and we also have a genetic basis for selection for what we know will survive there. Some of our ziziphus has died. Some of our ziziphus is doing just fine. So we are going to take seed from the ziziphus that we have growing there without irrigation. Among the plants that are doing the best are the *Moringa peregrina*, which are still fruiting on less than an inch [2.5 centimeters] of rain over the last two years."

Despite the suspension, Neal realized that the work had significance beyond this single, somewhat isolated village: "It's an agroforestry system we can implement anywhere in the west coast of Saudi Arabia."

When Neal remarked that policy changes set the stage for the pastoralists' plight, he was referring to modern history. But the story of expanding deserts certainly predates that. As authors like David Montgomery (*Dirt: The Erosion of Civilizations*) and Jared Diamond (*Collapse*) have written, the folly of mucking up our ecosystems did not begin in contemporary times. Nor is awareness of agriculture's destructive side limited to the present day. In 1864 and 1929, respectively, George Perkins Marsh (*Man and Nature: Or, Physical Geography as Modified by Human Action*) and J. Russell Smith (*Tree Crops: A Permanent Agriculture*) wrote about landscapes turned to dust as a result of wanton tree clearing for agriculture, particularly on slopes. In *Topsoil and Civilization*, Tom Dale and Vernon Gill Carter take readers on a worldwide tour of societal self-destruction. They write, "With the advent of civilized man, about six thousand years ago, the soil-building process was reversed in most areas where he resided: the quantity and quality of soil and the amount of life the soil supported all began to decline."[8]

With such an abysmal record of desert-proofing our landscapes, how can we be sure we've learned anything? In chapter 1 John Liu discussed the usefulness of "ecological forensics" to understand the unraveling of once-flourishing systems. This exercise points to practices to avoid—and toward ways to rebuild ecological function and reverse the destructive cycle. There are broad guidelines. According to John, the key is increasing biodiversity, biomass, and soil organic matter. As well as particulars, determined by place. Even the sober authors of *Topsoil and Civilization* grant, "Man has the knowledge and tools with which to destroy rapidly the soil and the plant and animal life it supports, but he also has the knowledge and tools with which to build soil and increase its productiveness much more rapidly than under natural processes. Man can apply his knowledge and skill toward soil building rather than soil destruction."[9]

Neal explores desert forensics in a webinar titled "The Problem with Agriculture" from the *Sustainable Design Masterclass* series, which he co-hosted. He begins by sketching out the typical food cultivation story arc: Farming starts in fertile floodplains where nutrient-rich soils have settled. Eventually the need for more farmland sends people to hillsides, where cutting down trees causes soil erosion and excess sedimentation in waterways. Trees are crucial to the hydrological cycle: stabilizing soil; cycling moisture within the local environment through transpiration; emitting minute particles, or aerosols—fungal spores, bacteria, pollen—that serve as precipitation nuclei and contribute to regular rainfall. The absence of tree cover leads to alternating floods and drought.

Irrigation, providing water between rains, creates its own problems, since water pools in low-lying areas and evaporates, leaving mineral salts. This results in salt-laden soil, a situation ameliorated only by flooding the area. Which in turn sparks more erosion. "This is the standard model of what happens to civilizations," Neal says. The sequence "generally takes three hundred to seven hundred years. It starts with agriculture."

The pattern was not lost on ancient observers. Neal quotes Plato's *Critias*, which narrates the demise of the mythical island kingdom Atlantis. Plato writes:

> The soil which has kept breaking away from the high .
> lands during these ages and these disasters . . . forms no
> pile of sediment worth mentioning, as in other regions,
> but keeps sliding away ceaselessly and disappearing in the
> deep. And . . . what now remains compared with what
> then existed is like the skeleton of a sick man, all the fat
> and soft earth having wasted away.

"Skeleton of a sick man" is an apt metaphor for a landscape that has lost all but the mineral scaffolding.

Land degradation and subsequent crop failures drove the Roman Empire's expansion, including Rome's forays to the Arabian Peninsula and North Africa. While it may be hard for us to imagine today, parts of what are now Jordan, Libya, and Yemen were thickly wooded. During the Roman conquest these forests were considered a military asset, as trees could be turned into ships, fuel, and housing. Neal points to the fate of Timgad, a Roman outpost in Algeria, whose buildings were buried beneath the expanding Sahara Desert sand, and therefore preserved: "The Romans cut down the forests to heat their baths and build up the city. They brought in animals to graze. Then they got massive soil erosion and a flood/drought cycle. Instead of water going into the ground, it flooded. It didn't rain as much because they didn't have forests. Timgad was sacked by Arabs in the 800s and totally abandoned. The conversion from a forested landscape to desolation was so complete that it was covered in sand."

For Neal, this broader history of ecological decline is the backdrop to the post-1952 story, which he keeps in mind as the Al Baydha Project strives to recover a capacity that's been lost.

Permaculture designer and teacher Rhamis Kent also works in arid regions: Egypt, Afghanistan, United Arab Emirates, Oman, Western Sahara, Palestine/Occupied West Bank, Somaliland, and Yemen. When your focus is addressing the problems linked with desertified landscapes, finding yourself in places prone to conflict would seem an occupational hazard: degraded arid landscapes lead to food insecurity, which then leads to political insecurity and instability. This has given Rhamis ample cause to ask how things got this way.

An American now living on England's southwest coast, Rhamis converted to Islam about twenty years ago and feels a special kinship with these predominantly Muslim areas. He initially trained as a mechanical engineer, and has a bent for the practical and for creative applications for tools. He once worked for Dean Kamen, inventor of the Segway, the tilting electric people-mover you see zipping around tourist zones. In 2008 Kent was running an Arabic-language program in California when he chanced to see *Greening the Desert*, the video of Geoff Lawton's Jordan project. "It pressed all my buttons," he recalls. This was engineering via nature's patterns and materials, bringing renewed life to a part of the world that had become important to him.

Toward the end of the video, Lawton said, "You can fix all the world's problems in a garden . . . You can solve all your pollution problems and all your supply line needs in a garden." What makes people insecure is that they don't know this, he said. Utopian as this sounded, it reflected the ideals conveyed through ancient texts Rhamis had been studying. He went to Australia to study with Lawton and is now—remotely from the U.K.—co-director of the Permaculture Research Institute in Australia.

I met Rhamis, an engaging black American in his mid-forties, at the 2017 Caux Dialogue on Land and Security in Switzerland. Within two minutes of exchanging hellos we were deep into a conversation about the environmental ethics expressed in Islamic philosophy and poetry. Five minutes later it was about how one might adapt used farm equipment to build water-holding trenches in Yemen. This is a guy who does things, and he's either halfway across the world solving problems or at home reading and studying to gain insights about solving problems—or at least an understanding of what exactly we need to solve. So I am not surprised when I later reach out to Rhamis to talk about deserts and he launches into a riff on his latest favorite read. Which at this moment happened to be a 2017 book called *Against the Grain: A Deep History of the Earliest States* by historian James C. Scott.

"He hits on how history has largely been driven by agriculture, and how agriculture was accomplished by four domestications: the domestication of plants, animals, fire, and people. A fifth could be water. These four, or perhaps five, things allowed for the population to be stationary. Scott says agriculture allowed for the establishment of states because grain could be stored and transported—parlayed into different kinds of wealth—and so the state could impose levies. People weren't necessarily chomping at the bit to stay on one piece of land. But agriculture requires steady labor, especially crops that require close care, as with the further domestication of varieties that became the bulk of our foodstuffs, cereals like oats, barley, wheat, and corn."

My head spins. To explain the expansion of deserts, I expected the usual sins of agriculture: removing tree cover, replacing perennial plants with annual crops, exposure of soil through tillage. I was not prepared to consider taxation, and how the quest to consolidate power led to settled populations, which resulted in the "forest— field—plow—desert" scenario.[10]

For Rhamis, an awareness of the cultural and ecological history of a place is always present: His mind is continually flipping backward and forward in time. Knowledge of the past can be helpful in designing restoration projects, he has found, as with his work in the Hadhramaut Valley in South Yemen, where in 2011 and 2013 he offered permaculture training in areas devastated by floods. Rhamis has a strong pull to this area, since it has long been a center of Islamic scholarship. "There are people in Tarim that have a direct connection to the prophet Muhammad and can run down their family tree to the prophet or the family of the prophet," he says. "It's as though you step back in time, because they are so firmly rooted in their heritage, in that way of comporting themselves."

For many of us, Yemen brings to mind the multi-year conflict, fueled by Saudi Arabia with support from the United States, that has led to starvation and suffering among the civilian population. In the shadow of this tragedy, Rhamis also sees complexity and richness. Yemen, he says, is "an epic place. It's very, very rocky. It's all escarpments, basically. As you fly over it you see nothing but rock. Then as you get closer you see valleys. Some of them are quite verdant, date palm groves and all these beautiful places that have been shaped by water and wind over time."

Due to its terrain, Yemen is marked by microclimates and variations in elevation, and is a place where many cultures have converged Numerous plant species have been carried in from elsewhere and taken root. This has resulted in a diversity of agricultural crops and farming traditions.

Like Neal, Rhamis seeks out older residents who may remember conditions and practices from decades past. While he was teaching a course in Wadi Hadhramaut, a student brought to class a book of South Arabian poetry, *Prose and Poetry from Hadhramawt*, translated into English by the twentieth-century Arabist Robert Bertram

Serjeant.[11] "It had poems that talked about what used to be planted in the landscapes in that region," says Rhamis. "The poetry ends up being a sort of botanical survey or animal survey for a given place—and a reference for a re-creation of these systems. Poetry is very much tied to the history of the Arabs. We were able to go back and look at these poems as a way of informing something that they might possibly do in the present."

The notion of a productive, fully realized garden as emblematic of paradise is central to Islam, says Rhamis, noting that detailed written descriptions of plant arrangements have inspired gardens through the centuries. The paradisiacal ideal "serves as a reminder of what one is aspiring to," he says. "The gardens of Al-Andalus in medieval Spain are considered the ultimate expression of what is outlined in the Qur'anic verse and in the poetry."

Even today, he says, the literature can serve as a useful template for arranging and assembling plants. He offers an example from the Greening the Desert Project: In 2010 Lawton was asked to consult on a farm at Wadi Rum, a red-sand plateau with dramatic rock formations in southern Jordan. It was a daunting prospect, because the land management had, to that point, been one of large-scale, high-input agriculture that withdrew water from an aquifer at an untenable rate. The soil was dry and lifeless.

Lawton crafted a design for the initial five-hectare (twelve-acre) plot that featured bands of mixed crops alternating with lanes of trees—in particular, leguminous trees that would bring in nutrients, block harsh winds, and provide shade. He drew a diagram and made a sketch of swales bordered by vines and trellises and shaded by a date palm canopy. It was a blueprint for what was possible once the soil improved.[12]

What Rhamis finds remarkable is how closely Lawton's illustration of swales amid rows of fruit trees—date, olive, fig, apricot, pomegranate, apple, pear—hewed to a well-known passage in the Qur'an, stating that those who believe and are righteous "will have gardens or perpetual residence" and that "beneath them rivers will flow." The verse continues with a story of two men: "We granted to one of them two gardens of grapevines and We bordered them with palm trees and placed between them [fields of] crops. Each of the two gardens produced its fruit and did not fall short thereof in anything. And We caused to gush forth within them a river."[13]

Which is essentially what Lawton did. His design was based on capturing water in the low-lying areas and distributing that to the fields and fruit trees, therefore seeming to create water by harvesting it. By 2013 the farm was reaping bumper yields of tomatoes, squash, eggplant, cucumbers, and many other vegetables and herbs.

Through literature Rhamis also came to understand that the archetypal oasis—that welcome and rarefied blend of scent, shade, and refreshment—is essentially a food forest. "When we talk about food forests, that's basically an agricultural oasis," he says. "It's a modern term for what people have been doing for a very long time. The oasis is a purposefully constructed assembly of ecological elements: You have places that are themselves very difficult to inhabit, and you're able to engineer a habitat based on your ability to put these things together. The idea is that if you understand the movement of water and where it tends to collect, you start from where there is water and work out from there."

Let's envision from antiquity an oasis, or composed botanical enclave, in the Hadhramaut area. From the historical record we could expect it to feature papaya, bitter orange, sweet lime, walnut, white or black mulberry, jujube, prickly pear; violet, lavender, mallow, wild thyme, chamomile, and rose; cypress, laurel, myrtle, sumac, caper, and sorghum, with all their associated fruit, seed, and floral elements according to season. Within each species, the options were vast. At different points there were more than three hundred varieties of dates in the region, and nearly eighty types of grapes.

What people chose to plant "wasn't just some kind of random thing," says Rhamis. "People knew what they were doing, and this was reflected in the poetic tradition." As an interesting parallel across the globe, researchers are finding that multispecies systems in the Amazon rainforest were not happenstance. Rather, these were intentionally created more than four thousand years ago by people conscious of the properties of different plants.[14]

Rhamis says knowledge of crop varieties and strategies for cultivation—for creating oases within ecologically and politically difficult terrain—are an important resource for refugees seeking sustenance and stability. In Yemen this is not an abstract concern. Nor did dislocation begin with the post–Arab Spring war. During flooding in 2008, thousands of people in and around Wadi Hadhramaut lost homes and most remain unsettled. Plus, despite the country's poverty, refugees steadily arrive from Somalia, Ethiopia, and Egypt. Certainly, migration is a significant regional challenge. Learning about agricultural tradition, Rhamis says, is a means for displaced people "to become reacquainted with their cultural heritage, the history of their people as related to the land."

In 2014 Rhamis and his colleagues were well on their way to establishing a regenerative agriculture project in the valley. With a budget and a design for a demonstration plot, they sent a team to Yemen. However, the Houthi insurgency started heating up. Though the violence was centered farther west nearer the capital Sana'a, the project had to be suspended. While Rhamis hopes to revive it, he concedes the difficulty of restoration in a war zone—even as war makes the task that much more urgent. "It's possible to go overland from Oman, but that takes a long time," he says. "Vehicles may sit at the border for days."

Rhamis's work shows the value of investigating past practices not just to fill in the time line of how things degrade—but also to gain insight into what works in a given system and how local traditions support that. This echoes John Liu's choice to study function as well as dysfunction, and Juergen Voegele's quest to find success stories amid the barren valleys. There is ecological wisdom all around us, once we open our eyes to it.

―――――――

By establishing a variety of trees and other plants, Neal Spackman and his Al Baydha team have made headway in creating a fertile enclave in the desert—a formidable feat. But Neal is not content to stop here. He believes it is possible to increase the rainfall outright.

During his podcast interview, Neal describes the geography of Al Baydha: "It is a microcosm of the rest of the region which in Saudi Arabia is called the Hejaz, which runs from Yemen to Jordan, all along the west coast of the Arabian Peninsula. We have the Red Sea and then a floodplain, and then there are mountains running north to south. So we have the hydrological dynamics of the sea."[15]

This topography creates a specific set of conditions, he says. Proximity to the Red Sea means there is a *maritime effect*: It is more temperate than farther from the coast, and there is abundant evaporation and humidity. The presence of mountains causes an *orographic effect*: Air is pushed to a higher elevation due to the shape of the land. As a result, he says, "You have a wet side of the mountain and a dry side of the mountain. When the humid air hits the mountains, it gets pushed up high into the atmosphere, where it cools and forms clouds and then it rains on the wet side of the mountains." By contrast, the presence of mountains blocks the dry side from receiving rain. The area remains in a rain shadow.

This setting between the maritime and orthographic effects means that Al Baydha is in a "closed-loop watershed," says Neal. "The water that evaporates from the Red Sea is the same water that comes as rain. That water isn't from a different system or from far away." The water can be continually recycled provided there's a vehicle for cycling it, as in plants. Because it is a closed system, he says, any changes to the hydrology will have only local effects: "It's not going to affect further downwind because any moisture hits the wall of the mountains and flows down the mountains back into the sea." The contained nature of the watershed, he says, creates opportunities to reinstate precipitation patterns in a land chronically deprived of consistent rains.

We know that moisture is always circulating, even in an area like Al Baydha where years might pass between rains. Why *doesn't* it rain in a desert? Neal highlights three possible reasons: (1) The area is in a rain shadow created by mountains. This is the orographic effect that prevents moisture from moving to the other side. An example is the Gobi Desert, which is in the rain shadow of the Himalayas. (2) Deforestation/loss of vegetation has led to a decrease in evapotranspiration. This results in drought/flood patterns and the loss of the small water cycle that would keep water circulating locally. This is the case in Al Baydha and the Hejaz region generally. (3) The aerosols present are not precipitation nuclei, and therefore have the effect of impeding rather than promoting the formation of clouds.

To clarify this last point: While airborne particles are always floating around in the atmosphere, they vary in how they function in the environment. "That same dust from the Gobi that causes rain in the Pacific Northwest may be a rain inhibitor in the desert," Neal says. "This is because when it's up in the atmosphere as dust, the sun reflects off that dust and heats the part of the atmosphere where rain clouds would otherwise form. The other aspect is that dust has a higher dew point, and so water won't condense on dust at warmer temperatures." According to Australian soil microbiologist Walter Jehne, dust particles wafting from eroding soil are too small for raindrops to condense around.[16]

Dust is considered inconsequential; the word itself conveys "something of no worth." Yet these specks of dirt floating about the globe have important implications for snowmelt, dispersal of nutrients or pathogens, and weather.[17] How does this play out in desert environments—and do those who work the land have any agency? Neal believes they do.

"We have massive amounts of dust because we've got all this bare ground. That inhibits rainfall for us. It's humid because of the Red Sea, but we don't have the cloud condensation nuclei needed for rain—and it's too hot. We've found that the rain coming off the floods is enough to reforest the area." Enhanced tree cover would dramatically reduce airborne dust, he says, "so you won't have that inhibitor anymore. Plus, you've got all the evaporation coming off your trees and you've got the sea with its evaporation. You've increased the density of water in the air."

Trees also emit cloud condensation nuclei, an assortment of pollens, turpenes, and organic compounds. And so by reforesting you have less dust, more ambient moisture, and aerosols that can produce condensation at higher temperatures: a scenario conducive to rain. Adds Neal: "You get this positive feedback loop because we're in this closed-loop watershed. We get more rain and then more plant cover and therefore more of those same effects so that eventually you've got a completely different climate."

The Al Baydha Project is not merely an exercise in making the desert bloom. In Israel, where people take pride in what they can produce in an arid environment, this has involved technology and precision irrigation to grow crops that may not actually be suited to the terrain. (There are exceptions, such as the Center for Sustainable Agriculture at the Arava Institute for Environmental Studies.[18]) Al Baydha's goal, by contrast, is a system that retains water and sustains moisture circulation. When they began, solar energy only generated heat, and the rain they got only led to floods. Now sunlight and precipitation support life.

In nature, life begets more life. As Neal says, it also alters the local climate dynamics. Which prompts a question about deserts and degraded lands the world over: How many microclimates does it take to make a new climate?

———————

In 2016 Ties Van der Hoeven—the Dutch engineer who joined John D. Liu on a return trip to the Loess Plateau—co-founded The Weather Makers, a company whose mission is to employ large-scale projects to restore the earth's climate systems through strategic interventions in coastal zones. This is an ambitious goal—especially since the first site the group seeks to re-green is a desert.

Ties's background is in a field that might seem the antithesis of regeneration: industrial dredging. And so we leave the realm of swaying date palms and fragrant herbs for less poetic backhoes, large-capacity barges, and pumps. Dredging is the digging up and moving around of sediment. In low-lying Netherlands and Belgium, dredging has long been used to turn flood-prone deltas into fertile farmland. The tale of the little boy who stuck his finger in the dike and saved his country epitomizes the extent to which waterworks looms large in Dutch culture. The Low Countries were alert to the rising of the seas long before its link to global climate change.

I also met Ties at the Caux Dialogue on Land and Security. Like many Dutch, he is slim and exceedingly tall. In manner, he radiates intelligence, impatience, and an air of nonconformity; he is confident in his thinking and will not be deterred by convention. "Growing up in Delft, I understood that this area was vulnerable. And that engineering was protecting our country—protecting against one-in-twelve-thousand-year storms," says Ties, now thirty-three. A sailor since childhood, he is a keen observer of weather patterns and waves. He studied hydraulic engineering at the Delft University of Technology, and began consulting in the field.

While on a contract with the Deme Group, a Belgium-based global dredging firm, he expressed concern that a project posed environmental risks. Word spread about this outspoken youngster who was upsetting veteran staff, and the CEO called him in. Undaunted, Ties said: "According to what I was taught, everything you are doing here would cause ecological damage. You could be in all kinds of lawsuits." The executive, mindful of the potential for litigation that haunts the earthmoving trade, asked: "What would you do?" Ties suggested the company stream data using sensors on the bottom of vessels to be vigilant about potential impacts. Deme took this step, which likely saved millions of dollars. Ties gained a reputation as the go-to guy for wild, innovative ideas—that might not always turn out to be so crazy.

———

In 2015 Deme was engaged in the Suez Canal expansion project, a multi-billion-dollar venture involving several large dredging firms. While in Egypt, the in-country manager, a friend of Ties's named Malik Boukebbous, learned about the plight of Lake Bardawil, once fertile and rich in biodiversity and now extremely shallow and saline. Separated from the Mediterranean Sea by a narrow sandbar, the lake remains an important spawning area for fish. As eroding soil has washed in over time, the lake, once some thirty meters (one hundred feet) deep, has dwindled to a depth of one and a half meters (five feet). Overfishing, including bottom trawling, has led to ecological problems, notably high mortality among sea turtles. Local people asked Malik if the dredging company could do anything. Malik brought this question to Ties.

"I said, well, yes. We've got to increase tidal prisms," Ties recalls. A *tidal prism* is the change in the quantity of water between low and high tides, that which flows and ebbs in an inlet or bay. Increasing the tidal prism would enhance Lake Bardawil's dynamism, thereby bolstering life in the lake and potentially beyond. The idea would be to work at the edges of a system in order to reboot the hydrological cycle and enrich life in the watershed, thereby creating the conditions to rehabilitate the land. Ties was thinking like a dredger. Altering tidal prisms is something a dredging enterprise would be uniquely equipped to do.

The notion of intervening at the level of tidal prisms suggested dredging's potential for eco-restoration. "This was the first time I got a little sparkling in my head," Ties says. This gleam of an idea led to a period of intense study, during which Ties would wake up at 4 AM to pore over diagrams. He puzzled questions like: *What reduction in salinity would enable the fish population to rise?* and *How much vegetation, starting with sea grasses, reeds, and algae, is needed to capture moisture from the sea and mountain breezes?*

Like Rhamis Kent and Neal Spackman, he dove into ecological history, zeroing in on thousands of years ago when the Sinai was vegetated. He learned that throughout this period, winds from the Indian Ocean brought moist air to the entire region of which the Sinai is at the center: the Middle East, North Africa, and Mediterranean. And that some four to seven thousand years back, many parts of the world experienced severe weather disruptions—often the alternate flooding and dry spells that characterize desertification. Climate turbulence shows up in literature and religious narratives: the epic deluges and droughts of the Old Testament; the Mesopotamian flood in *Gilgamesh*. Ties theorized these events were the result of a change in wind direction: rather than receiving moisture, the Mediterranean/ North Africa region was now exporting it. With the help of research based on core sediment sampling and pollen studies he confirmed that the winds had indeed changed.

The question was, *Why?* Whether natural climate shifts caused land to deteriorate or the land degradation prompted climate extremes remains a mystery. But Ties has a theory. He believes that the stripping of vegetation in the once-verdant cradle of civilization altered wind patterns and, subsequently, the hydrological cycle. From his knowledge of sailing and dredging and examining both old trade maps and current weather maps, he inferred that agricultural land use in ancient Egypt altered wind direction and wind speed. Which then disrupted the flow of moisture.

Ties says that with its mountains and location, the Sinai represents the continental divide of an entire region: a juncture that marks where rivers and streams flow toward different water systems. In consequence, a change in conditions here could disrupt moisture distribution throughout. This led him to ask: *If the Sinai's land degradation contributed to a disturbed regional climate, could its rehabilitation help reinstate functional weather systems?* "We know from the history that there used to be more wind in the Nile Valley, as this enabled transport vessels to go up the river," he says. "This gave us our first understanding that there are acupuncture points for the globe."

Ties studied wind depths and weather patterns and concluded that regenerating the Bardawil watershed could modify wind direction and rain patterns. "Sinai is a perfectly shaped triangle form and has areas of elevation, including Mount Sinai and Mount Catherine. So when the sun heats up, it tends to function as a vacuum, pulling air upward. The air cools at higher elevations and becomes more dense, which creates a vacuum pump to bring in the moisture."

He homed in on three factors: (1) a change in the local moisture regime; (2) the Sinai's topography; (3) and the Coriolis effect, which describes the movement of air or objects in the context of the earth's rotation. From this he determined that revegetating the area around Lake Bardawil would change the local weather. And that the moisture once lost to the Indian Ocean would potentially return to the region.

While this may seem wildly speculative, there is growing scientific evidence that vegetation does bear on wind activity and precipitation. For instance, Western Australia's rabbit-proof fence, put up in the early 1900s, did little to deter vermin but offers a great land-use experiment: Along more than six hundred kilometers (375 miles) we can compare acreage cleared for farming with natural vegetation, separated only by a wire barrier. Studies found that the natural woodland area received more rain than the side with annual crops. A 2002 paper describing this phenomenon by the late Tom Lyons bears the evocative title, "Clouds Prefer Native Vegetation." Modifying the vegetation, he wrote, "leads to significant changes in land surface characteristics, such as albedo, surface roughness and canopy resistance. These land surface changes induce changes in the atmospheric boundary layer."[19] (This last term refers to the one-kilometer-thick wedge of sky that's in contact with the earth.)

Further, research indicates that temperature differentials between uncovered land and vegetated land can influence air movement and even cyclonic storms.[20] The Belgium-based nonprofit WeForest emphasizes the role of plants, especially trees, in transporting moisture and highlights the concept of the *precipitation shed*: the sourcing area for rain.[21] Professor Millán Millán, an advisory board member of The Weather Makers, writes about the moisture concentrations required for rain and the conditions that enable this level of humidity —namely adequate vegetation to provide moisture through transpiration.[22]

Like Neal's premise that active management can return rain to western Saudi Arabia, Ties's plan has significant implications, as well as some significant "ifs." But as was the case with the Loess Plateau, understanding the factors that led to collapse can point to restoration efforts that may reverse that process. History offers us examples of biome shifts, in which, say, a forest turns to savanna or a savanna becomes desert. (Indeed, we are seeing this happen around the world, although it's often labeled drought or famine.) Neal and Ties are exploring ways to move ecologies toward enhanced ecological function: to bring about increases in biodiversity, biomass, and soil organic matter. Their work poses a crucial question: *What threshold of land restoration—what size or density—brings about a state change in a positive, more vibrant direction?*

As for the Sinai Peninsula, Ties says its long-vanished waterways are still with us, literally written into the land. After thousands of years of soil loss, he says, "the pedosphere [earth's soil layer] has been removed and it's down to the lithosphere, or geological material. This allows you to look via satellite and see the old river systems: the hydrological cycles of the past etched in stone." While we're on a video call, Ties shows me a slide of the Sinai that depicts its ancient river system. It looks like a heart, crisscrossed by linear veins and arteries.

———————

How do you bring life to a watershed now reduced to its geological bones? With organic matter, according to Ties. "The lake has been collecting solar energy for thousands of years," he says. This energy is embodied in the very sediment—soil, sand, and silt from the highlands—that now threatens to smother the life out of Lake Bardawil. Ties believes that this stored energy, this battery of accumulated carbon stocks, can be used to revive the ecology. And that this will bolster the marine ecosystem and enhance the well-being of local communities, people whose livelihoods are threatened while the water grows murky and the region dries out. This exemplifies the principle that what is considered waste can also be seen as a resource: It all depends on the context.

As a dredging engineer, Ties is a student of sediment. He understands that the muck isn't in itself a problem; the trouble is that it's residing in the waterways as opposed to the coastal marshlands. Such deposits are rich in nutrients and carbon, both essential to thriving ecosystems. Ties, Malik, and their advisers arrived at the concept of Resource Based Dredging: using dredging technology for large-scale eco-restoration. Central to the process is increasing the flow of energy, organic material, and marine life between the lake and the sea—amplifying the tidal force—to enhance ecological vitality. Dredging inlets would increase tidal prisms, dilute salinity (thereby improving water quality), and increase oxygen and currents to stir up sediment. This will ensure the nutrients are available to plant and aquatic life. "There is more than a billion cubic meters [3.5 billion cubic feet] of organic material in the lake, and organic matter is the kick-starter of every ecosystem," Ties says. He adds that the tools and precision of dredging will enable them to construct saltwater marshes for birds and other wildlife.

Ties goes on to explain that more plant life means more evapotranspiration, which means more solar energy dispersed (cooler temperatures) and more water cycling in the local environment (more rain). With more plants, carbon and nutrients like nitrogen and phosphorus can help to build biomass, rather than becoming pollutants in the water or the air; again, a substance can be a resource or a toxin, depending on location and concentration. Greater complexity also means more surface area for gas and moisture exchange and the formation of microhabitats. "Vegetation creates shadow, and shadow is cooler so you have more available moisture and more condensation," he adds.

All this is intended to spark beneficial feedbacks: increasing biodiversity, sequestering greenhouse gases, and increasing groundwater levels. Then there's the fish, the mullet and bream that are the economic base of the region. Ties says the fish glide along the currents to seek food, algae, and phytoplankton that would be stimulated by precision dredging.

By the end of 2016, Ties's project attracted interest from a number of people at the forefront of ecological restoration, several of whom serve on The Weather Makers' advisory board, including John Liu; Spanish meteorologist Millán Millán; biologist and designer John Todd; and Australian paleontologist Tim Flannery, whose 2001 book on anthropogenic climate change, *The Weather Makers*, inspired the company name. Personally, I find the prospect of large dredging vessels cruising around fragile watersheds a bit unsettling. There is the potential for things to go awry, or for unintended consequences. A dredging project to expand the Port of Miami, for example, destroyed more than half a million corals, including threatened staghorn coral, as they were smothered by plumes of sediment that the dredging stirred up.[23]

As a tech-shy romantic I would prefer for ecological vitality to emerge through an unmediated alliance with nature. In Farmer Managed Natural Regeneration (FMNR), for instance, dryland areas are reforested by cultivating the shoots that grow from felled tree stumps, thus nurturing an ecological process already under way. In 2018 Tony Rinaudo of World Vision received a Right Livelihood Award for FMNR work, as did farmer Yacouba Sawadogo of Burkina Faso, who revived local forests by building on traditional water-saving methods.[24] As another example, in Holistic Management grazing ruminants are the driving force behind large-scale land restoration.

Many restoration strategies involve machinery, however, to break up compacted soil or redirect the flow of water. We have tools that, while potentially harmful, can be applied cautiously and productively. As for dredging, Ties says it's a matter of changing the intention of the industry. When it comes to damage, he advises being careful about assigning blame. Often risks escalate because a client doesn't want to spend the money to do it right. The destructive side of dredging, he says, is "part of the destructive movement of humanity. Everything has to be cheap. Let's not blame the industry. It's us that created this industry." Knowing what we know, he says, entire sectors, like industrial dredging, can shift their orientation toward ecosystem regeneration. Ties's point is that we can all be part of this change.

Rhamis is offering a permaculture design course with Maged El-Said at the Habiba Community in Nuweiba, which El-Said co-founded in the South Sinai. The hope is that Bedouin in the Sinai will take part in the training and be on board with the work in Lake Bardawil, which lies midway between Port Said and the Gaza Strip. Recent unrest in the peninsula, particularly in the north, is a challenge and adds a layer of uncertainty.

Back in Al Baydha, Neal Spackman says he's seen more indications of ecological recovery. Such as the influx of fungi: "The number of species of mushrooms has quintupled. Used to be that I'd see one species of mushroom after a rain, and there were just a couple of them.

Now when it rains, five species pop up—and there are more of them. What that says to me is that the mycelial web is developing. That's a foundational thing for us because mycelia release water as part of their respiration. Marginally this is increasing water in the soil."

He says that via their extensive hyphae networks, the mycelia are scouting out sand and silicate particles to deliver nutrients that the trees need in a bioavailable form. Neal showed pictures of their fungi to renowned fungi expert Paul Stamets. "He said that is this species, this is that species, and all of this means that you're going through ecological succession. This means we are moving from a desert to something that is not a desert, in an actively managed way."

Insects have also increased. "There used to be only flies and mosquitoes," Neal says. "Now we have crickets and ants. You don't think of ants as a thing that isn't there. The first time I saw ants was 2015. And I had looked for them the whole time. One ant colony on a hundred-acre farm is not going to build an inch of topsoil, but the fact that they're there now and they weren't before indicates that we're moving in the right direction." They also now have termites, which he calls "the earthworms of the desert. Termites take the lignin in the wood and digest it and move all that stuff underground."

The number of birds and nests has grown, as has that of snakes and lizards. The group built a pigeon house in 2011, and without stocking it with food or water, some twenty pigeons—"a staple desert garden animal, kind of like the backyard chicken for the desert"—have claimed the spot. "We could be managing them much more intensively and have a couple of hundred of them," he says. "As a barometer it's very interesting. The fact is that our ecology is developing and pigeons have moved in. The same with bats and our bat house."

In 2019 Neal completed an MS in business at Stanford. "Now that we know what we can grow, for me the next step is creating viable enterprises," he says. He and his colleagues in Saudi Arabia have lots of ideas. They can use mesquite to make gluten-free flour, for example: "Good for diabetes, which is a problem in Saudi Arabia." Fruit and leaf powders are used in Chinese medicine, which would guarantee a market, he says, adding that in other cultures "jujube honey is considered medicinal. It is one of the few food crops mentioned in the Qur'an." He and Stanford classmates have recently founded Regenerative Resources, a company that combines shrimp aquaculture with mangrove agroforestry and arid agroforestry on desertified coasts. Their first site is in Baja California Sur.

Neal also sees possibilities for silvopasture, which combines grazing and trees, for Al Baydha. "There's the potential to do dairy, meat, and leather off these grazing systems," he says. "Eventually we get to the point where we are not buying feed to feed the animals, so it becomes a low-input system with products we can sell at the local market. Especially during Haj." He points out that seven and a half million pilgrims are expected to visit Mecca over the next three years. That puts them a half-hour drive away from the site, he says, and "all those people have to eat."

Meanwhile, Neal says the project has yielded a "living prototype and template for the afforestation of Saudi Arabia's Red Sea Coast." The vegetation is "growing and producing now without management, without irrigation, without any inputs whatsoever—including twenty-eight months with no rain—and will continue to develop as time goes on." This shows the degree to which ecology can revive under even the most inhospitable conditions, and how this provides hope for regions marked by the twin scourges of land degradation and rural poverty.

— THREE —

Beyond the Impossible

Conflict and Consensus in New Mexico

*We do the impossible all the time—we just
don't recognize it.*

—Bob Chadwick

The Rio Puerco, a tributary of the Rio Grande, wends a
north-to-south path through the center of New Mexico. It
has the dubious distinction of bearing the highest sediment
load of all US waterways, and third highest in the world. The name
means "dirty river" or, literally, "river of pigs." Along Rio Puerco's
eroded banks—about 120 kilometers (seventy-five miles) north of
Albuquerque, well past the sprawl of Bernalillo—lies the town of
Cabezon, named for the volcanic peak that looms over the broad,
flat terrain. Never a large settlement, Cabezon was a stagecoach stop
and farming center in the late 1800s. By the 1940s this part of the
Rio Puerco had dried up, and at decade's end Cabezon's post office
had shuttered. The Catholic church Iglesia de San José, a putty-
colored adobe topped by a bell tower and wooden cross, is among
the few buildings still standing and intact. It offers a Mass once
every few years.

Descendants of the ranchers who once flourished here still reside
in the vicinity or come up on weekends from Bernalillo. With the
exception of Navajo and pueblo lands, the surrounding area is under
the jurisdiction of the Bureau of Land Management (BLM). By
arrangement, several families still keep cattle on their allotments.

One boon for the community is that, dry though the landscape is, there is an artesian spring and therefore adequate water: water for ranchers and water for townspeople, handled by respective overseeing boards.

Ironically, this good fortune has become a source of contention, as the two groups and individuals within them vie for control. A rash of escalating conflicts—including gunfire and sabotage at the pump house—have erupted and BLM staff are summoned, sometimes multiple times a week, to calm things down. Even in ordinary times this stark rangeland is regarded as a lawless area. Now, with neighbors turning against one another, the situation has grown dire. As Tim, who raises cattle there, puts it, "My children are afraid to go to the property because of the strife. People are going around armed."[1]

A seemingly intractable conflict over resources. A landscape degraded over time. A community that has lost its way.

This is why I am here. I want to find out: Can these people get past differences and turn their attention to healing the land? If so, this will offer important lessons for others struggling with tense disagreements and damaged land—and who may not see that the two are connected.

As a nonfiction writer, I've considered knowledge the crucible for change. I trusted that once people had a command of the facts and applied these toward deeper understanding, they would do the right thing. Sure, history has served up plenty of instances of stupidity and hubris. But we know more now, and isn't the point of knowledge to ensure we don't repeat our mistakes?

Imparting information remained my orientation to writing as I explored and learned: about the multilayered, often surprising benefits of living soil; about how climate is influenced by water and water is mobilized by plants; about the connections and synergies among living things. I took it as a given that if I were articulate enough—if I could craft the most elegant phrases and illuminating metaphors—readers would inevitably embrace productive solutions to ecological problems on their own and their communities' behalf. Surprising as it may be, I've held on to this blithe confidence in the value of information for a long time. "If only people understood . . ." I would say to myself, kept going by my certainty of the power of knowledge.

I have now lost that belief in the primacy of fact. I won't take space here to discuss the larger social and political context of a crumbling trust in basic reason, though I'm sure that's a large part of it. For me, the awareness that information alone is not enough hit me like a revelation. Disillusioning as it has been, this understanding points toward other routes to change—pathways that could prove more transformative than "knowing" on an intellectual level.

In September 2017 I was in Mexico for a four-day meeting of the Regeneration International network at the Vía Orgánica ranch, a research and training center near San Miguel de Allende linked with the Organic Consumers Association. It is a beautiful setting: We convened under high, airy tents from where we could gaze out at fields full of cosmos—a flower native to Mexico—in all shades of pink, and follow the dips and turns of bees and butterflies. There were more than a hundred of us from around the globe; I was among colleagues and friends. Yet I was restless, my attention drifting as I sat in the sessions. Running in the back of my mind, like the hum of those hovering bees, was the dawning realization that all the knowledge and technology needed to shift to a regenerative future —one marked by agriculture that builds soil carbon, retains water, produces nutrient-dense food, and revives land and communities—is already available. It's only people that get in the way.

In the middle of listening to a panel discussion on promoting regenerative agriculture worldwide, I decided to take a break. I wandered toward the exit and noticed Jeff Goebel and his wife, Myrna, sitting in the back of the room. I had heard about Jeff from Peter Donovan of the Soil Carbon Coalition. He knew Jeff well and was keen on his consensus work. Peter told me about the consensus model, calling it the only approach he knew that produced lasting personal change. Also, that it involves many hours or days of talking in a circle, conscious listening, and building trust.

"That sounds interesting," I'd said to Peter in a noncommittal way, thoroughly dismissing the idea. I was too impatient for that sort of thing. Besides, at that point, I believed that achieving meaningful change was simply a matter of getting information out. By late 2017 I was no longer so sure.

Kneeling down and careful to speak quietly, I inquired about the Community Consensus Institute. By way of explanation Jeff told a story about work he did in the late 1990s in Mali, in the same general region along the Inner Niger Delta that John Liu visited in 2010. The community was in crisis. Eighty-five percent of the residents faced food insecurity, and the seven ethnic groups that lived or passed through seasonally were at war with one another. Particularly violent were land conflicts between farmers and pastoralists. For this USAID-funded Sustainable Agriculture and Natural Resource Management (SANREM) project, Jeff was hired to hold a series of workshops that included members of various tribes.

"During the second workshop, I asked what people wanted and they said they wanted to increase food production," he said. "I asked if they thought they could increase it 10 percent without Western technology. They said sure. I asked if they could manage a 20 percent increase. They looked at each other and, tentatively, said, 'If we work real smart and real hard, we could probably do that.' Then I asked if they could do 50 percent and they all said, 'No way!' I asked, Why not? They gave me all kinds of reasons: People are lazy, the soil is no good, not enough money, not enough rain."

With that settled, he said to them: "Given that increasing 50 percent is impossible, *if it were possible* what would you do?" People came up with several ideas: They would diversify crops, plant legu-minous trees, implement grazing strategies that build fertility, and so on. They saw that there was actually a lot they could do.

Fifteen months later Jeff returned to Mali. "I was in the vehicle driving up to the meeting site," he said, "and people were lined up

along the road waiting for me. They could hardly wait to show me—
they had increased food production 78 percent!"

This was quite a story. "How was this possible?" I asked.

"It was their set of beliefs that convinced them it was impossible.
Once we acknowledged it couldn't happen, it took the pressure off.
This freed up their minds so they could consider what they *would do.*
Hey—do you have time right after lunch? We can do a brief version
of the workshop."

I set out to find more partakers and brought in Renald Flores, a
former investment banker from France who specializes in soil restoration and holistic decision making. He wears dreadlocks, perhaps to
put distance between his current life and the years he spent in a suit.
I'd met Renald at Caux, where he gave presentations on culture and
decision making, including in the corporate world. "I am going to be
very honest about the banking culture," he'd said, and then proceeded
to paint a sordid picture of the nihilism, greed, and competitiveness
that festers beneath the bright, confident handshakes and silken
neckties. I was ready to plead with him: *No more honesty, please!*

The four of us reconvened in a quiet corner near the main building and went around the group responding to a series of questions,
switching the speaking order each time. The first query was, "How do
you perceive humanity's relationship to the natural world, and how
do you feel about it?" Surprise: None of us was especially happy about
the current state. The next was, "What is your greatest fear connected
with this situation?"

Fear? I had been doing everything possible to fend off any scary
thought related to the environment, warding away threats to the inner
barrier protecting me from a grief I imagined would be unbearable.
Fear was a hard, lacquered knot buried deep in my psyche. I trusted
that as long as I didn't trouble it, this nub of fear would remain inert.
This is one reason I held tight to my focus on solutions. I wanted
to reassure myself of the potential for environmental regeneration:
opportunities to ally with nature that will be available to us when we
as a society are ready to adopt them. How could I possibly articulate
my fear, given the lengths I'd gone to convince myself it didn't exist?

Still, consensus is about trust. And so, cautiously, each of us went to this dark place: acknowledging fear that the world would become uninhabitable, that humanity could go extinct, that our children's children would subsist in a lifeless world of concrete. This was painful to say and to hear, even as we knew our words could only hint at the depth of the dread we harbored: apprehension of an existential nature, usually unspoken and therefore untouched. At least in this moment, however, none of us was bearing it alone.

Fortunately, we didn't leave it there. The following question was, "What would be the *best* possible outcome?" This is something we rarely consider, perhaps because managing fear absorbs so much of our energy. We shared what we wished for: For climate change to provide an opportunity for people to come together. For nations to stop spending money on war and instead devote money to healing the planet. For people to appreciate that we are part of nature. For 50 percent of global agriculture to become regenerative within three years.

Then the same two follow-up questions that Jeff posed in Mali: "What are all the reasons this is impossible?" As in Mali, our responses were resigned shoulder shrugs: a variety of reasons, mainly related to politics and money. And finally, "Given that our wish is impossible, *if it were possible* what potential actions and strategies could help bring this about?"

That simple phrase "if it were possible" proved a potent invitation, and we launched into high-gear brainstorming. At one point, Renald said, "I would knock on the door of conventional farming companies, and I would say: 'Let's work together. We've seen how your way works, now try ours. Give us three years. If you don't like the results, you can go back to what you were doing.' I would go straight to Monsanto's offices and say this." (Gasp! Monsanto?) "We all need each other," Renald continued. "Let's give others the chance to be part of it."

I don't recall the other ideas, but I remember how it felt: breaking through the trepidation barrier—putting words to what had been locked away, isolating us—generated a sense of release. And a sense of energy, previously sapped by the dread that lurked beneath our actions and assumptions.

———————

It was out of desperation that the Bureau of Land Management office in Albuquerque hired Jeff. In recent months staff was continually catering to emergencies in Cabezon. None of the usual conflict-resolution measures worked. Someone in New Mexico recommended the Community Consensus Institute and, out of options, the BLM called Jeff. A three-day workshop scheduled for March had to be postponed because several of the ranchers were due in court—not a surprise since ranchers, townspeople, and the BLM were all suing one another. It was rescheduled for early April.

Jeff and Myrna, who live just south of Albuquerque, pick me up the day prior and we drive to the Jemez Senior Center, the workshop site, where a Zumba class is under way. We get settled at our lodgings, Cañon del Rio in Jemez Springs, a bed-and-breakfast/spa with nice grounds and a hot tub. It's midweek, so only one of the town's two sit-down restaurants is open. Over chiles rellenos and bottled beer, Jeff clues me in to what to expect.

"The venue is afraid there will be fights and people throwing chairs out the window," he says, noting that this recently happened with this group. "There will likely be law enforcement there." He is dismayed about the workshop's delay, since he'd already interviewed several people in the community; a lapse between interviews and the workshop slows momentum, increasing the likelihood that some attendees will fall away.

"We'll work with who's there," he says. "I've had cases where the people having the conflict weren't present—and the conflict ended up being solved." The lawsuits cast a shadow on the process, he adds, because they create suspicion, including that people will use statements they make in the workshop against one another in court. Generations-old resentments are another factor. Some families have been in the area for hundreds of years, with long legacies of extra-marital affairs, broken marriages, and feuds.

Jeff is in his sixties, of medium height and stocky build. He has a calm, deliberate way of talking that belies his intellect; there is a kind of physical stillness about him. Myrna is petite and radiates competence. In Guam, where she is from, she founded and ran a multi-million-dollar landscaping business. The two have been together just a few years, beginning soon after Myrna, fresh from a divorce, arrived in New Mexico at her son's urging with little more than two suitcases. She is committed to Jeff's work and handles many of the logistics, including food.

"My worst possible outcome is that they can't let go of past grudges," Jeff says. "My best possible outcome is that all the energy that's been taken up in conflict can be devoted to creating healthy land. That is my intention: to shift that energy to do something good for their community and the earth."

———————

We arrive at the center well before 8 AM, when the workshop is set to begin. The parking lot is mostly empty. A tall, older man stands beside a large truck. He introduces himself as Roger and indicates the lack of vehicles. "I'm the only rancher here. The troublemakers aren't here. Why would they want to be? They just want to do what they want to do." I'm not sure which troublemakers he's referring to. I smile and say I look forward to seeing him in the workshop, then go inside. There I find Jeff, who is arranging twenty-five chairs in a circle. He says, "It's important to have more people than chairs. If not, it casts doubt on the value of what we're doing." He's counting on more than twenty-five participants, which, at this moment, seems ambitious. People stroll in slowly. I meet BLM employees and representatives of the state's livestock board: people whose job it is to be there.

I sit down and silently will people to walk in. I want to believe this will work. A group of four comes in, three men and a woman. Then a slightly built, athletic guy with a shy smile. A few more people arrive and just one or two empty chairs remain. I glance around the circle and see felted western hats and baseball caps. Leather cowboy boots and rubber-soled work boots. A lot of man-spreads.

Jeff initiates the "grounding" exercise: people introduce themselves and share what they're hoping to achieve. "One goal is to meet my neighbors," a slim, soft-spoken older man named Guy says. "Hi, Roger," he adds, leaning in to catch the other's eye. "We never see each other!" A tall older man wearing a gray hat and yellow work gloves ambles over. He's not walking well. He finds a chair and climbs up to perch on the backrest, leaning forward to favor his back. We're halfway around the circle when the door swings open and a large, very tall fellow strides over to the last unclaimed seat. He wears a smooth, brown, wide-brimmed hat and a silver belt buckle the size of a small dinner plate. People shift their chairs to enlarge the circle, though there's plenty of space. He seems not to notice their shuffling and settles into his seat.

Jeff waits a beat before continuing. "We only see a part of the world. People who have opposite views each have part of the solution." He uses the example that when we're in a circle, everyone sees a different slice of the room. I'm looking toward the kitchen entrance, at the coffee urn and pastries Myrna has set out; the man with the bad back sitting across from me would see the large bingo board. "My hope is that we can see the world as bigger than we do now by acknowledging others' perspectives," he says.

Jeff refers to the Native American legend of the "coyote trickster": creating mischief, in order to impart a lesson. He says, "The gift is the other person, to enable us to see what we can't otherwise see. If you don't learn the first time, the coyote will keep appearing until you do." We talk about the nature of conflict. Jeff quotes his mentor, Bob Chadwick: "Bob used to say, 'Conflict is my friend.' That's because it creates opportunities to learn."

We divide into groups and record responses to Jeff's pre-chosen process questions on a flip chart word for word, for the simple act of accurate recording builds trust. It is a sign that one has been heard. A sheriff in uniform strides in and says, wryly, "I'm here to make sure you don't kick each other's asses." He paces the back of the room, hand on his pistol, while we ask questions, respond, and record.

Noon-ish, we break for lunch. I bring mine outside to take in some spring sunlight. I join a table right as someone is saying, "It's the two brothers' fault. If it weren't for them, we'd be fine." A snicker makes its way around the table: the gratifying bond of shared disdain. Who are the brothers? I think of the big man with the shiny belt buckle, whose name I've learned is Frank. We return to our groups and flip charts.

I can't lie: This is getting tedious. We are talking around the conflict. Like everyone else, I've been dutifully answering questions about how I feel about conflict and change and being listened to (or not). I've taken my turn facilitating and recording people's responses in broad colored markers while resisting the temptation to edit everyone's words. And I am getting impatient. Earlier, Jeff talked about impatience: "We never have time to do things right. But we have time to do them over and over." True enough: How often do we learn from our mistakes? I take a deep breath and decide to roll with it.

The next round inquires about the worst possible outcome of *not* confronting conflict. It turns out that avoiding conflict poses the same threats as engaging with it: more violence, more costly lawsuits, and as one woman says, "the feeling that I could get a flat tire on the road and my neighbor won't stop to help." No one foresaw this similarity.

"The fear of confronting conflict paralyzes us," Jeff says. "Some people are already living their worst possible outcomes—but fear makes it difficult to change. Ideally we can shift from fears to hopes, because what we focus on tends to manifest. The idea is to shift to what you want to see."

Jeff emphasizes that at any given moment, worst and best possible outcomes are both present, and that we can choose. Worst possible outcomes tend to dominate our thinking because that's what we're wired for: It's the fight-or-flight response that activates our adrenal system. (Jeff expresses this as a verb: Something "adrenalizes" us.) Confronting fears defuses their power, so we can focus on best possible outcomes and the beliefs, behaviors, strategies, and actions to bring them about. We return to our large circle for end-of-day reflections. I notice that whereas earlier there had been veiled references to conflict and the "two sides," this seems to have ebbed. Instead, people talk about jokes and lighthearted moments they'd shared in their groups.

Before dinner in Bernalillo, Jeff, Myrna, and I pile into Jeff's car and head out to see the area. Except for the Cabezon summit in the distance, the land is flat and you can see a long way. The desertification is extreme, although the stark, mineral-ly terrain has a kind of beauty when it catches the light. We stop and look at the Rio Puerco. Several feet down past the brittle riverbanks carved out by erosion, a bit of water pools on the mud.

"Look at all the water that has run off the land. It blows my mind," Jeff says. "This was once the granary of New Mexico, rich river-bottom land. They used to call it the River of Cottonwoods. There were beavers here, and bluebirds. One of my dreams has been to restore this area. You've got to think like the grass, to imagine what it could be." I reflect on John Liu's comment that the state of a given landscape mirrors the consciousness of those who dwell there. Perhaps it goes both ways: A landscape that's suffering gives rise to people who are suffering. And troubled people create troubled landscapes. Would the conflicts plaguing the community have emerged amid beavers, bluebirds, and rich, fertile soil?

We drive through the town of San Luis, whose residents rely on the domestic water. There's the Lonesome Dove bar and a few trailers. Cabezon, a few miles farther, has the church and some old adobe and stone relics. Jeff points toward where the pump house is. "There are a few dozen grazing leases in the area," he says. "They do continuous grazing, and now there's no grass. People are going into other people's allotments to graze."

One trespass led to gunfire. According to the version Jeff heard, the bullet went over the targeted man's head and landed in a pickup truck. "This is what people talk about when they say 'tragedy of the commons.' It's an attitude of scarcity," he says. Rather than leaving the animals to graze down to the ground, he says, ranchers would be better off with planned grazing, actively moving cattle so that the plants have a chance to recover.

On our way out, we see a single, forlorn-looking cow. Jeff sighs. "They're not creating habitat for the future, but covering up the sins of the past. That's ensuring the problems will happen again."

Jeff was raised and spent much of his life in the Pacific Northwest, where his father taught range management at the university level. Jeff, too, pursued land management and conservation and worked with agencies like the USDA and US Department of the Interior. Over time he observed that projects often stalled despite best intentions, and began to explore consensus work. "I saw great plans that sit on the shelf and nothing happens," he says. "I didn't want that. I wanted results. What makes things happen is a change in behavior, and that happens when belief systems change." He defines *consensus* as "100 percent agreement to do the right thing." He distinguishes this from *compromise*, which—though generally considered the ideal—means that everybody gives up something.

He has been inspired by the late Bob Chadwick, who developed consensus models for businesses, school systems, and land and resource conflicts. From all I've heard, Chadwick was a wise man, near sagelike in his trust of others and ability to suspend judgment. There is a brief film of Chadwick in Nic Askew's Soul Biographies series in which he describes a consensus program in the Klamath River Basin in Oregon where tensions over water had sparked violence. He brought together farmers, ranchers, tribal leaders, and environmentalists to create a common vision. Chadwick spoke about the human need to connect and be listened to. What's important, he says, is that after a workshop "every person feels they have been seen."[2]

Jeff knew Bob Chadwick when he was a child. Bob and his wife, Carolyn, were in the same social circle as Jeff's parents, and the two couples often went to local dances together. Jeff recalls that when he was ten, the Chadwicks had six children younger than he was and Carolyn was pregnant with their seventh child. That winter, one of the boys accidentally started a fire and the family lost everything. "Our family pitched in to help," he says. "It was odd for me to see the four girls wearing mine and my brother's boys' jeans." Two months afterward, when Carolyn went to the hospital to give birth, she suffered a blood clot and died.

While Jeff lost track of the Chadwicks over time, he says, "I can still remember, from before these incidents, seeing Bob's love for his wife on his face. I could just tell he loved this woman dearly. That's why this had such an impact. He had lost everything: his house, his wife, and had seven children to raise, including a new baby daughter. Yet this is someone who says, 'Conflict makes us the best we can become.'"

Years later, Jeff was working with the Confederated Tribes of the Colville Reservation when someone suggested Bob Chadwick join the project. He'd said, "I knew a Bob Chadwick when I was a kid." Indeed, it was the same one.

Jeff says of Bob, "He made an effort to seek richness, even when nothing looks good. He strove to see the world in joy and awe and curiosity and abundance, to say, 'Today I'm okay and this is all that matters, because none of us has tomorrow.'" One time Jeff took part in one of Bob's workshops in Oregon. The region was in drought. "Bob said, 'It's going to rain because of you all coming together.' I went up to him and said, 'Why did you do that? It was going so well, and now you've made this ridiculous claim!' He simply went on with the workshop and within a half hour it starts to rain."

Uncanny events seem to follow consensus work, Jeff tells me. For example, the Mali SANREM project consisted of five workshops. Each took place during the dry season, and every time it rained. "One time we were in a school with a tin roof," he says. "The rain was so loud it felt like the sky fell down. For twenty minutes no one could hear anything." The second time this happened, people in the group started chattering among themselves with looks of surprise. The interpreter told Jeff they were saying, "He brings magic to this place." One participant from the Fulani tribe who had not borne a child for five years became pregnant around the time of the workshop. "She was telling the village women that my work creates fertility," he says, still amused years later. "She named the child Jeff—and in this world where boys are named Mohammed and Ali!"

———————

Since the first day of the Cabezon workshop had gone well, we hoped word would spread. Yet on day two, Friday, it seems we are fewer, as several agency staffers are committed elsewhere. "I always believe that whoever is there is the right people to be there," Jeff says more than once. This leaves me skeptical, and privately I wonder if this will be a flop. At 8 AM on the dot, Jeff launches the session with a riff about power. "Power tends to equalize," he says. "If people resent power held over them, they will find a way to undermine it, often unconsciously."

After our greeting circle we take up the question: "What is the situation in Cabezon and how do you feel about it?" One person new to the workshop is the director of BLM's Albuquerque District Office, Danita Burns. She makes it clear that she means business. "Most managers would have said, 'We're pulling everything up,'" she begins. "I'm not going anywhere. New Mexico is special because people here are really connected to the land. You people are related by blood and culture. As for the landscape, it looks like crap. It's been beaten up on. We need you to fix it. Cabezon is worth saving. But we need guidance. Our last-ditch effort is to bring Jeff here. Let's get on it and make it work."

Her summary helps fill in missing pieces for me. Decades back, the BLM had helped set up a water storage and delivery system to serve the ranching allotments, as is its purview. The cattlemen chose to extend the water line to domestic users, and a separate board was created. The head of cattlemen's group, the majordomo, governs access to the pump house. This is currently Frank, the large man with the silver buckle I noticed the day prior. When a dead bird was found in the pump house and the domestic board was cited and fined, people suspected shenanigans from Frank. The water board installed surveillance cameras and someone came in and turned them off. And then things got ugly.

Others weigh in. "The bottleneck is the pump house, and that's because of the majordomo. Clear and simple," says Teddy, the slim, soft-spoken older man. He represents the domestic group, which, for reasons I still fail to understand, is suing both the ranchers *and* themselves, legal maneuvers that are costing tens of thousands of dollars.

"I don't have trust in the cattle association," says Tim, who wears a baseball cap. He has an allotment but is allied with the domestic group because he depends on the water when he's in town. This seems to be a dig at Frank. "Cattle are dying. I don't feel anyone is representing me."

George, the rancher with the bad back, interjects, "What is the problem? I don't see a problem."

Tension in the room is rising. Conflict, tidily contained yesterday, is bursting out. All the while Frank, the majordomo at the center of the storm, sits in his metal utility chair, quiet and impassive, betraying neither anger nor hurt.

Jeff raises the topic of change. "Even if you don't like the way things are, there is a sense that the devil you know is better than the devil you don't know," he says. "When you do accept the need for change, it's a psychological double whammy: There is the grief at what you once had and is no more, *and* the stress of the unknown future."

Time for small groups. We scoot our chairs to our respective stations and write at the top of the flip chart, "What is the worst possible outcome of not changing?" After we acknowledge that change is needed—because the cost of not changing is unacceptably high—we explore: "What do we want the change to look like?" And: "What would be the evidence that meaningful change is happening?"

I reflect on board meetings I've taken part in and realize that we rarely devote time to a collective vision. Typically, the emphasis is on deciding something, which often comes down to who presents a case with the most conviction. You ride Robert's Rules through the voting process and you've done your job. But decision makers will be more accountable for the results when they establish agreement on a collective vision, rather than simply moving through a checklist.

In Jeff's workshop, everyone shares the same objective: "We'll know we've made a change when we stop fighting each other."

On toward 5 PM we return to our large circle. Burns, the district director, says she's been moved by the workshop. "I knew you cared for the land but I hadn't seen commitment," she says. "You weren't there yet. Now I think you're there. If there's a fault, it's mine. I believe we've all been acting from our own filters—for this is how we were raised to see the world. Now the question is: What's stopping everyone from moving forward?" She pledges to offer whatever support she can. Cabezon and its notable peak is an icon for this part of the state, she says. "It's on the poster in my office."

———

Jeff talks a lot about how the mind works. In sessions he is deliberate about engaging both right and left sides of the brain—considered the realms, respectively, of emotion and cognition. His typical opening question, some variant of "What is the situation and how do you feel about it?," is designed to tap both sides. He says the human brain is a powerful problem-solving tool: It is primed to solve problems, and so perceives situations from a problem-solution standpoint. This creates a tendency to define goals according to problems to be solved, an orientation that limits our ability to visualize what we want. What we regard as "the problem" is frequently a symptom of an underlying or systemic problem, a reality that a problem-solving approach may blind us to.

The crucible of this model is the shift between "worst possible outcomes" and "best possible outcomes," that deliverance from despair to possibility. Getting there involves challenging the built-in tendencies of our own minds. Worst possible outcomes command our attention because our brains are signaling *danger*. This awakens the mechanisms that process fear, a response many times faster than conscious thought. Worst possible outcomes are based on our own past experiences and therefore feel real, even inescapable. This can become a self-fulfilling prophecy. When people believe the worst will transpire, it becomes more likely because their behavior aligns with that outcome.

"People often say they're 'pragmatic,'" says Jeff. "That's code for accepting worst outcomes as the best we can do."

There's that mischievous brain again: It thinks it's doing us a favor by steeling us for a bad situation, when in doing so it actually steers us headlong toward it.

I often think about these dynamics as I read and listen to the news. Articles and broadcast segments are increasingly framed as: "Why Climate Change Will Make [fill-in-the-blank] Worse," or "How [some proposed policy] May Make Bad Things Happen." Headlines are written to rivet our attention—you get the most clicks when you "adrenalize"—and they seize our attention because that's how our brains work. Worst-case-scenario perseverating keeps everyone pumping stress hormones. This leaves us at once frozen and over-whelmed, unable to act or seek alternatives. In short, we stuff our heads with worst possible outcomes and wonder why we get them.

It is tough to think clearly amid a surge of adrenaline. One priority in consensus is getting participants to slow down and listen—which allows you to focus on what you desire rather than what you fear. This wished-for future is "intensely imagined and strongly felt," says Jeff, even if not drawn from personal experience. "When expressing best possible outcomes, you often people hear people laugh. And say things like, 'Wouldn't it be cool if . . . ?' You come up with solutions you hadn't thought of before." As with negative outcomes, believing them affects a person's perceptions, beliefs, values, and strategies; belief in positive outcomes "tends to be self-fulfilling . . . when strongly held."

That is what I remember so vividly from our mini workshop in Mexico: that sense of inner release; an unshackling that let me shrug off fear, or at least get some distance from it; a sense that deep hopes and aspirations, hardly articulated even to myself, were within grasp.

Community Consensus Institute is among many approaches people are using to address conflict, make decisions, and manage change. There is also sociocracy, sometimes referred to as dynamic gover-nance, which emphasizes inclusion, democratic values, and personal autonomy. Another model is Theory U, which describes itself as "Leading from the Future as It Emerges." The program has evolved through several decades of work and research at MIT, particularly that of Otto Scharmer, and focuses on decision making through individual self-awareness and presence with one another.

What unites these models is an invitation to bring one's whole self into group processes. They also provide alternatives to the standard meeting format, which many people find disempowering due to time constraints, hierarchies, and the tendency for the same people to hog the megaphone.

Then there is Holistic Management (HM), developed and popularized by Allan Savory. An early adopter, Jeff has been a HM trainer since the mid-1980s. Holistic Management is not, as some people think, simply a strategy for managing livestock in the service of grassland restoration. It is actually a decision-making framework that can be applied to any management task, from running a farm or ranch to starting a business to crafting public policy. It happens to be ideally suited to the planning and implementation of grazing—fair enough, since the challenge of overseeing the multiple moving parts of grazing systems is what prompted Allan Savory to devise the model. Savory's writings and presentations increasingly emphasize the need to shift from reductionist thought, which limits the capacity to manage complexity, to decision making based on a holistic context that clarifies and articulates values, needs, and goals.

After practicing Holistic Management in his work with the Natural Resources Conservation Service (NRCS), BLM, and USDA Forest Service, Jeff came to feel that, helpful as the model was, it hardly made a dent in the fears, lack of trust, and reluctance to change that left promising plans languishing on the shelf. He started incorporating consensus in the 1990s, a period when he was Integrated Resource Management Planning (IRMP) coordinator with the Colville Confederated Tribes in eastern Washington.

When new in this role he invited Chadwick to meet with a group of forty people addressing forest management. He recalls, "I told Bob, 'Take whatever topic you want—anything but clear-cutting.'" This was too volatile to touch. "So Bob does the initial interviews and says, 'We're going to focus on clear-cutting.'"

Two months later Chadwick asked Jeff to arrange a meeting in the forest next to a proposed clear-cut. Jeff invited everyone to walk through the forest in solitude. He then posed the question: "What can we learn from the clear-cut that will help the tribe be successful?" The Elders said, "In our culture, you don't take everything." The foresters said, "There are a lot of diseased trees. We need to clear it."

The logger, a tribe member said, "I understand what the Elders say. I understand what the forester says. There are five or six large pumpkin pines per acre in there. Those are worth about $5,000 per tree." He saw that by keeping the more mature trees, they could cull diseased trees, maintain the essence of the forest, and avoid the erosion, flooding, and other problems associated with clear-cuts. He could leave $25,000 per acre on the ground and say, "I don't need that"; he could hear the different sides and tune in to his own beliefs. Says Jeff: "That day is when I realized the power of Bob's work. It allows for a new way of seeing the world."

The Colville tribes then held about half a million hectares (1.3 million acres). Despite a constant parade of expert consultants, conditions had deteriorated. Springs and streams were drying up; problem insects increased while wildlife numbers declined. Jeff observed group meetings and noticed they were dominated by white people: the planners and technical experts accorded professional status. "The tribal members were often quiet, and public meetings had few tribal members." He figured their attitude was, *They don't listen to us, so why go?*

Another project involved producing a watershed plan for Six Mile Springs, an area of six thousand hectares (fifteen thousand acres). Jeff approached the two non-tribal project managers and said, "Let's go out with the Elders." He instructed them to let the twelve Elders do all the directing—to listen and not dominate the conversation. Jeff recalls: "We drove around and went where the Elders wanted to go. When we got to a lunch stop I asked the tribal leaders, 'How do you feel about the management of the reservation today?' It got heavy. They said they felt shame, disgust, hopelessness. They said, 'We're the Elders. We're supposed to be providing for the next generation and we feel powerless.'"

He said, "'With this watershed we have an opportunity to do something different. What would this land need to look like for you to feel pride?' They said they'd want the pumpkin pine back.

Migrating wildlife. Medicinal plants that used to be there. Streams flowing again, not just intermittently. Jobs for the people." He turned to the two white men and said, "You guys just got your marching orders." The forester initially protested, "I didn't go to forestry school to grow medicinal plants!" Jeff said to the native Elders, "We're not going to do anything until we come to you first and say: This is what we heard you say."

The plan took a year and a half for the team to develop. It considered the Elders' priorities, including medicinal plants and the overall health of the forest. Jeff presented the proposal to the Elders and said, "'Here's what we thought we heard you say. How did we do?' The room got really quiet. There were tears. The Elders said: 'We have never been asked or listened to before.' It passed unanimously, then went to the community and the tribal business council."

Consensus helped the tribe identify specific goals and thereby develop plan components to meet those goals—say, the number of trees to cut or how many miles of road to pave. They were actually able to harvest 50 percent more trees because they realized through the decision-making process that it would be in the interest of forest health. They realized that they only needed seven miles of new road, as opposed to the twenty-one miles they'd initially projected. "The Forest Service had budgeted $125 an acre—and typically spent $300," says Jeff. "We budgeted $75 an acre and the cost turned out to be $25 an acre." In the third year they trimmed $16 million from a $55 million budget without cutting any jobs—and delivered the budget three months early.

The third and last day of the workshop is Saturday. There were more concerns about violence; someone had tipped off the sheriff that people were planning to disrupt the meeting. Some new people show up, including Bruce, the young man who leads the domestic water board, but others drop away. I sense in Jeff a weary resignation. He sits quietly, knees spread, his back curved and shoulders raised like a football player at rest.

Jeff begins with the story of a workshop in Montana. Ranchers despaired that their children were leaving because they saw no future for themselves in livestock. The consensus process revealed that they knew better herd management—specifically, holistic planned grazing —would increase their earnings, but they worried their neighbors would look askance at them if they suddenly started doing things differently than everyone around them. Now face-to-face with the neighbors whose judgment they feared, they realized how silly this was; their neighbors had the same fears.

"Taking care of land isn't an economic problem. It's a people problem," Jeff says. "I don't want a fix here in Cabezon to be a short-term fix. I want it to be long term. People are tired of fighting and suing each other. You've got water. For now. Out on the land I see sheet erosion. That soil is your future storage for rain. You can change management so that you can keep water on the land, and bounce back quicker after a drought. The choice is: Do you want to dwell in drought and scarcity, or create a new future? If land is in poor condition, 10 percent of the rain goes into the ground. If it's in good condition, *90 percent* goes into the ground. That's good for plants, healthier for animals, and the whole cycle changes. I want to help you turn this around."

I see Jeff has brought ecology into the conversation. Until now it had remained in the background. The attitude had been, *Let's fix the conflict and then everything will be fine.* I am curious to see where this will go.

This being the final day, the goal is to come up with a collective statement: a compilation of participants' visions for the future, a guide and reference point for the community. For the moment, however, Jeff says it's time to address the conflict at the heart of it all.

He asks Frank and Bruce—leaders of the ranchers and domestic water district, respectively—to move their chairs to the center, and for each to choose a "listener" whose role is to reflect back what is said. This is crucial since misconceptions, so frequent during tense exchanges, can derail the process. Bruce picks Martha Graham of the New Mexico Source Water Protection Program and Frank chooses Marianne, whose husband is George, the older fellow with the bad back.

The encounter assumes its own rhythm. Bruce and Frank ask each other questions and respond, maintaining eye contact as Martha and Marianne confirm and paraphrase each reply. The hall is as silent as it's been over the three days. Through the men's body language, a parallel narrative emerges: Bruce crosses his arms; Frank leans back; Bruce drops his hands to his lap; Frank sits up with a jolt.

Each articulates his frustrations. Frank describes the indignity of having multiple cameras beaming down on him every time he enters the pump house, which he's responsible for. "There was a video sent out about me. Why?" he asks. "Charges were attempted via Homeland Security. I was investigated for two hours and then cleared. But I wasn't cleared enough for the domestic side." He pauses and admits that he was so enraged he took a broom and slammed the cameras. This is what had been caught on video.

Bruce looks concerned. He folds his hands. "I needed to report the possible contamination to our compliance officer," he says. "I was told we could face charges. The cattlemen's association is a nonprofit. We're a quasi-governmental entity. We have a lot of oversight." He explains this is why they needed the cameras, and why the video remains accessible to the public.

They share worst possible outcomes. Frank's is that the cattle group will continue to suffer because of the accusations against him. Bruce's is that the fighting persists. Frank commiserates. "The worst is the families," he says. "Your daddy and I were friends in high school." As for best possible outcomes, Frank's is restored access to the pump house, and for people to understand that he doesn't intend "malice, destruction, or, most of all, sabotage." He emphasizes that his is the first lot on the water line, so he has a stake in the system. Bruce's wish is that people hold to an agreement. "We all have to follow procedure for public safety," he says. "Finances get audited constantly. We are under a microscope."

Asked what he is willing to do to resolve the situation, Frank says, "I am willing to be helpful. I see that we took it a little—a lot—too

far. My part is to keep my mouth shut and keep going." He refers
to a saying in Spanish: "If I can't help you, I sure won't do you no
wrong." He says he is willing to step down as majordomo if it's best
for the community.

Jeff says, "We may have a new majordomo, but it might still be
Frank. This is because Frank will be different. And his behavior
change will have an impact on the community."

A BLM officer frequently called to Cabezon to put out flares
thanks the men and the two listeners and says, "When you guys go
back, there will be naysayers. But they weren't here—you were."

Danita Burns says: "You are my heroes. I appreciate the role of the
listeners. It can be hard to see someone you're in conflict with. The
important thing is: You have been witnessed."

During lunch Frank and Tim, a rancher who's taken the domestic side of the quarrel and kept his distance from Frank, are sitting
together, amiably chatting.

After the break, Jeff shows a brief film about Ute Creek Cattle
Company, a fifty-six-hundred-hectare (fourteen-thousand-acre)
ranch in northeast New Mexico that has improved its land through
better grazing management. The creek, previously dry and overrun
by salt cedar, now flows year-round. With the return of wildlife,
the property has a bird sanctuary and is a popular destination for
photographers. I wonder what the group makes of the story, but no
one says much.

We devote the rest of the afternoon to the collective statement.
In preparation, I had taken the best possible outcome pages and cut
individual comments into separate paper strips. We assemble these to
create a narrative of change.

That the morning's face-off had an impact is reflected in people's
comments as we work. One rancher says, "I can do a better job with
the grazing."

George, the rancher with the bad back, says, "I need to manage my
anger better."

Oliver, Frank's brother, says, "I'm going to speak differently about the community."

Marianne says, "I see the BLM differently. I used to see them as 'the government.'"

Owning personal responsibility seems to be contagious.

The collective statement reads: "We are looking forward to getting work done on the ground and people are excited about the future and about bringing kids to the land. We are building a model to emulate."

———————

I leave New Mexico invigorated: I felt I'd witnessed something special, a true shift. People had let down their guard, and in doing so glimpsed an alternative to living in fear and suspicion. The sight of Frank and Tim eating lunch together was emblematic of the change. I caught others staring wide-eyed, or nudging each other with elbows: If these guys could make amends, maybe anything was possible.

But what about the long term? Even if people leave their guns at home and smile at each other more, it won't mean much if the land continues to degrade. If they are fighting about the pump house when they have ample water, what will it look like if and when water no longer flows so freely?

As it turned out, a lot did change in the weeks and months following Jeff's workshop.

Frank was replaced as majordomo of the Cabezon pump house. He was disappointed but accepts this. There are improvement projects in the works, including collaborations with the Bureau of Land Management, which has continued to regard Jeff as a resource for this region. People are working together and the atmosphere, once so tense, has eased. They have started having community get-togethers again.

Through much of 2018 Jeff used consensus to build a coalition to improve soil health and create momentum for legislation that would address flooding, drought, and other challenges related to land degradation. Early in 2019 the New Mexico state legislature over-whelmingly approved the Healthy Soils Act, which was endorsed by two hundred groups in the state. Its mission is "to promote and support farming and ranching systems and other forms of land management that increase soil organic matter, aggregate stability, microbiology and water retention to improve the health, yield and profitability of the soils of the state." Within one year, this went from being an idea shared by a group of colleagues to state law with funds to disburse. The legislation is a model for other states, and creates a framework for educational programs in schools and at on-farm workshops.

Jeff has been asked to introduce the consensus-based approach to help inform state legislative water policy. He has also been asked to bring regenerative agriculture into the discussion of a proposed New Mexico state banking system, so that state funds can be invested locally. (State and other public banks are mandated to serve the public interest as opposed to private shareholders.) One of the largest landowners in the state has invited Jeff to consult on improving his hundred-thousand-hectare (250,000-acre) property.

There is also growing interest in holistic planned grazing among Cabezon ranchers. Jeff is now organizing training opportunities. One rancher who took part in the workshop developed a grazing plan for some of the acreage, which has been approved by the BLM and the New Mexico State Lands Office. There is movement toward restoring the Rio Puerco through improved grazing management. Ranchers who attended the workshop have organized programs like a field day dedicated to regenerative ranching on the Rio Puerco.

While this is a result Jeff was hoping for, regenerative grazing was only briefly touched on during the three-day workshop. Jeff alluded to the erosion, and the fact that by squandering soil the community was losing its capacity to hold water, but he didn't directly introduce these topics for discussion and consensus goals. The film he'd shown about the ranch at Ute Creek seemed tossed into the schedule as an afterthought, and few people commented about it afterward. Apparently, it made a big impression on the group, however. Several ranchers later asked Jeff about it, inquiring about how they could improve their grazing management. Maybe this was the point, strategically midwifed by Jeff: The de-escalation of tension created a mood that allowed people to be open to new ideas and to shift from accusing one another and defending themselves to improving the landscape they all share.

Meanwhile, the conversation I had with Jeff, Myrna, and Renald Flores at the Regeneration International meeting in Mexico is bearing fruit in surprising, if indirect, ways. Early in 2018 Dalmas Tiampati, a young Maasai herder who founded the Maasai Center for Regenerative Pastoralism, reached out to me. Dalmas was concerned that drought and land degradation in Kenya were threatening his tribe's way of life. He had introduced holistic planned grazing to his community but felt training and education alone weren't enough.

I told Dalmas about the Community Consensus Institute, and soon several of us—Dalmas, Jeff, Renald Flores, Precious Phiri, Africa director of Regeneration International based in Zimbabwe, and Karen Wagner, a colleague of Jeff's in Oregon—were regularly meeting on Zoom to discuss how Dalmas might initiate a consensus process in his region. Through the video connection, I see Dalmas in his village with a child on his lap, Renald in the south of France, and Karen sitting in the dark (the call starts at 6 AM Pacific Time). Behind Jeff there is always the same painting—a southwestern vista of a river, tipis, and mountains topped with snow. We haven't yet figured out how to fund the consensus project, but Dalmas has initiated conversations with NGOs and begun raising awareness of consensus as a tool.

Renald Flores, who is busy managing land restoration projects in Sweden, Mexico, Turkey, and India, calls our meeting in Mexico a turning point for him. "When I said to bring about regenerative agriculture I would go to the big guys' table and ask if we could work together—that was a deep moment for me." And now it is happening. The manager of one of Latvia's largest farms learned about soil regeneration research Renald had done, approached him, and asked if they could work together to improve their productivity. The opportunity to work with, and have impact on, conventional agriculture was exactly what Renald envisioned when he talked about being at the big guys' table, and giving everyone a chance to be part of the shift to regenerative agriculture and ecological restoration.

"The important thing is to understand the farmer's mind-set," Renald says. "It's not blaming the use of chemicals but understanding, *Why do they do it?*" He says reasons may include a focus on farm operations, including the costly machinery and inputs, or a lack of understanding of the soil. "It's not that people want to use glyphosate. They just don't know of a viable alternative," he says. Renald is setting up research plots with compost, cover cropping, and eliminating tillage. The manager says if organic methods reduce input costs while maintaining yields, he'll transition the whole farm.

Precious is integrating consensus into her facilitation work. She recently conducted a training with safari operators and community members on the fraught topic of land access for grazing. Safari owners expressed concern about clients, who pay steep fees to view wild animals, seeing people walking around leading cows wearing bells. "When I asked about the worst possible outcomes, I could see the stress levels go up," says Precious. "Then when we looked at best possible outcomes, the safari people were able to understand that holistic grazing can bring back rivers and wildlife. The land is just crying for help." Rather than opposing grazing, the safari owners saw there were reasons to embrace it.

Of course, it remains to be seen what happens in Cabezon in the long run. However, what the community has already accomplished offers lessons for other towns, villages, tribes, and groups beset by grievances, inertia, confusion, and assumptions about what is or isn't possible. Fact sheets or step-by-step guides are not the answer. Nor are appeals to fear; we have too many layers of internal defenses for that. Not to mention that fearing one another is what leads to community paralysis in the first place.

The experience shows that even in highly contentious settings there are ways forward, if we can acknowledge and respect our shared humanity. Part of that humanity is the desire to be heard and

understood. Feeling heard dispels tension, leaving us all more receptive to new ideas. It is hard to listen or be curious when you're scared,
suspicious, or pumped with adrenaline. Engaging in a way that builds
trust helps us all become more open to opportunities, seeing not just
barren land but the possibility of oases.

FOUR

They Belong to This Land

The Reindeer Chronicles

Although it is well-known that animals are influenced by climate, an interesting question is the extent to which animals can themselves influence climate.

—"Megafauna and Ecosystem
Function from the Pleistocene
to the Anthropocene"

Through most of 2017, the court rulings were tilting in the young man's favor, despite the odds against him. Jovsset Ánte Sara, a then twenty-three-year-old Sámi reindeer herder in the far northern Arctic highlands, had challenged the Norway government's order to reduce his animal numbers to seventy-five. Jovsset Ánte said such a drastic cull would preclude his ability to continue herding, but his refusal to comply wasn't simply a response to his own plight. As he told the *New York Times*, "I sued because I could not accept to see my culture die."[1]

Two lower courts affirmed the herder's right to keep his animals. However, the government continued to appeal and at the end of 2017 the country's Supreme Court determined that Sara must adhere to reindeer quotas or face additional fines and other penalties, including the forced slaughter of his herd, for which he would have to bear the cost. The young man is bringing the case to the UN Human Rights Committee, claiming that the Norwegian state's actions infringe on the rights of indigenous citizens. The government was not willing to await the UN's conclusion before imposing the slaughter, Máret Ánne Sara, Jovsset Ánte's older sister, wrote to me. With the support of the East Finnmark District Court, a September 2019 date was announced. So that the animals wouldn't be killed, Sara gave them to a relative. "The temporary emergency solution offers protection . . . until April 2020," Máret Ánne wrote. "What happens after is still uncertain."

Like every incident that captures the popular imagination, the herder's case has elements that make for good drama. Jovsset Ánte is attractive and youthful. In news clips he stands resolute, even regal, in traditional Sámi attire, the *gákti*: dark blue wool with colorful striped trimming along the neck, shoulders, and chest. Plus, the confrontation involves reindeer, that beloved hoofed mammal of the far north, with its grand, felted antlers and cozy holiday associations.

In an ironic twist, that spring Norway's vaunted "Slow TV"—an unlikely phenomenon that has allowed viewers to binge-watch knitting and firewood gathering in real time—featured the annual reindeer migration: seven days of airtime for a thousand-plus animals to make their way across the tundra from winter pastures on the plateau to summer grazing near the coast. State broadcasting service NRK was apparently unaware that among the herd moving majestically across the plains were reindeer at the center of Jovsset Ánte's lawsuit against the government.

The young herder's case has also remained in the public eye because of his sister, an internationally known artist who regards the government's punitive actions as an assault on her community. Máret Ánne Sara's *Pile o' Sápmi* project began with the 2016 installation of two hundred reindeer heads heaped in front of the Finnmark District Court where Jovsset Ánte was to be tried.[2] The work alludes to "Pile of Bones," a name the Cree nation gave to land that became Regina, Saskatchewan, as a means of retaining their connection to the buffalo their community had always hunted. It refers also to iconic nineteenth-century photos from the United States depicting trophy buffalo heads piled as high as buildings. The link to these images highlights the scale of the slaughter that nearly wiped out the once-ubiquitous animals.

Máret Ánne's goal is to call attention to the impact of severing indigenous communities from the traditions and wildlife that anchor their identities. In her view, the government's imposed limits make it impossible for young Sámi herders to continue as their ancestors have done. An established herder with more animals could better withstand a proportional cull than someone like her brother, who is just starting out and buying equipment and already working within narrow margins. "If you look at it from this perspective, [it's] an effective way to end the reindeer herding if our young people can't do it."[3]

Máret Ánne has fine, blond hair and looks equally stylish in artworld chic and traditional Sámi dress or a combination of both. She says it gives her strength to wear traditional clothes, often handmade by family members, and wears a silver amulet, which is said to offer protection.[4] She has a soft, gentle voice, but conveys resolve. She expresses concern that the pressure on Sámi herders is coming from a government that many people consider relatively progressive. She describes this as a new colonialism, in which the infringement on the autonomy and land access of indigenous people is often more subtle than it has been historically, but equally destructive.

"Norway is one of the most fair countries in the world, one of the countries that is most known to respect human rights," she says. "We have the only official indigenous parliament in the world." And yet, she says, the Sámi parliament's involvement can be deceptive, because its only mandate is a consulting role. Since the government isn't obliged to follow any recommendations from the Sámi parliament, sometimes it amounts to little more than theater.

In December 2017 Máret Ánne installed four hundred reindeer skulls shot through with bullet holes and strung up with wire to create a wall-sized curtain in front of the Norwegian parliament, as Jovsset Ánte was appearing before Norway's Supreme Court to defend his right to keep his herd intact. In this third trial the young man lost his case. The State disregarded the expressed wishes of the Sámi parliament. Máret Ánne's artwork has since been acquired by Norway's National Gallery.

Of the approximately sixty thousand Sámi people in Norway, only about 5 percent herd reindeer. Still, the vocation remains a significant part of the cultural heritage. "Sámi reindeer herding is, for me, the Sámi bank: for language, handicraft, knowledge of the environment, ecology," Máret Ánne says in a film.[5] The government claims that current reindeer numbers must be curtailed to minimize damage to the fragile tundra ecosystem. She says this belief, shared by most Norwegians, is "so simplified and polarized that it cannot by any means justify such drastic punishments on people, animals, and society."

I learned about Jovsset Ánte's case in May 2017 when I was in Norway to speak at Kunsthall Trondheim, a contemporary art gallery, at a symposium titled "Indigenous Knowledge: The Practice of Sustainable Existence." The event was timed to commemorate the one hundredth anniversary of the first national assembly of Sámi people, which took place in Trondheim in 1917.

My topic was the pastoral tradition and ecology of grazing. One of the most interesting aspects of this subject—and the source of much confusion and disagreement—is how many grazing animals a landscape can sustain. It would seem to make sense that the more animals, the greater the impact, and most people believe this to be the case. But nature doesn't work this way. Building on the insights of French farmer and scientist André Voisin, Allan Savory has shown that it's not the *number* of animals that leads to overgrazing but rather the *time* plants are exposed to grazing pressure.

For example, if cattle hang around the same spot indefinitely—say, by a riverbank—they may damage it, whereas two or three times the number of cattle kept on the move might benefit the land. This makes sense intuitively. Under wild conditions, ruminants would never stay anywhere long; predators would be at their hooves. Nor is animal impact necessarily negative. Grassland ecosystems co-evolved with ruminants, whose actions upon the land stimulate important ecological processes. Savory came to realize this in the early 1960s as a young game ranger in southern Africa. He observed that when people fenced animals away from deteriorating land, the land's condition got worse instead of reviving.

I spent a week with Allan Savory at the Africa Centre for Holistic Management in Zimbabwe and saw the extent to which over a period of fifteen years holistic planned grazing helped this seasonably arid landscape rebound. The Dimbangombe River runs a kilometer (two-thirds of a mile) farther into the catchment, and has been flowing throughout the year.[6] Where there had been bare soil, grasses grow thick and abundant. And biodiversity has flourished, with larger herds of wild species like sable antelope and the return of wetland birds. In fact, the Africa Centre's biggest challenge has been acquiring enough ruminants to keep up with forage production.

But what about reindeer? That is to say, do the Norwegian government's actions make good ecological sense?

Reindeer keep their landscapes good and cold, which is significant because northern climates are among the most rapidly warming landscapes on earth and melting permafrost threatens to release vast stores of methane and CO_2 into the atmosphere. A research team led by Mariska te Beest of Umeå University in Sweden found that reindeer browsing and feeding on shrubs during the summer keeps plant growth under control. This is important because shrubs and small trees have a lower albedo, or reflectivity, than the grassy heath that would otherwise dominate. The darker-colored bushy plants tend to absorb solar energy, therefore accelerating thawing. By contrast, the heath reflects more radiation and so does not take in that extra heat, keeping the area cooler. According to te Beest and her colleagues, the effect is likely limited to areas of high reindeer density.[7] So culling them could undermine this positive effect.

The Nordic Centre of Excellence (NCoE) Tundra also notes the climate-positive impact of reindeer in a warming world. Grazing is a bulwark against "shrubification" in the region, which proceeds as warmer temperatures create conditions favorable to more woody plants. "By preventing the invasion of trees, tall shrubs and forbs, reindeer maintain the openness of the tundra, which is a precondition for the survival of many small-sized arctic plant species," a NCoE Tundra report concludes.[8]

Reindeer can also help maintain permafrost by crushing the snowpack with their hooves, according to Sergey and Nikita Zimov, father-and-son research scientists in Russia. The Zimovs developed a project in the late 1980s, in which they brought herbivores that thrive in arctic conditions—reindeer, moose, Yakutian horse, bison, musk ox, yak, Kalmykian cow, and sheep—to their North Siberia reserve. The goal of Pleistocene Park, as they call it, is to re-create the productive Mammoth Steppe ecosystem that predated human expansion into far northern latitudes.[9] The blanket of snow that cloaks the tundra for much of the year acts as an insulator, and this protects the soil surface from the cold, Nikita Zimov explains in a 2017 interview with *PRI's Living on Earth*. "When animals trample down the snow, they actually thin that layer of snow, making it dense, and this allows much deeper freezing during the winter."[10] This sustained chill can extend snow cover to the spring months, which means maintaining higher surface albedo longer into the year. It also keeps the permafrost frosty, so that the microbial life in frozen soil doesn't activate and consume organic matter, a process that releases greenhouse gases. In an experiment that compared areas with and without herbivores, the Zimovs found that soil temperature in places where animals grazed was lower by at least 15°C (27°F).[11]

Other research suggests that under warming conditions, heavier grazing in the tundra is associated with increased carbon fixation—since grass grows more quickly than brush and carries more enhanced photosynthetic capacity—which could compensate for anticipated carbon losses.[12] Still more studies point to heavy reindeer grazing resulting in improved nutrient cycling and production of biomass.[13]

And so the policy developed by the Norwegian government in order to "protect the environment" may have it all wrong.

———

My guide to Norway is Ulf Ullring, a biologist who manages the Langsua National Park near Lillehammer, which spans more than five hundred square kilometers (two hundred square miles) of birch woodland, old-growth coniferous forest, alpine heath, and wetlands. Reindeer certainly come through the park, but moose predominate. Visitors to the area can arrange for a "moose safari" to watch the King of the Forest in its element. Ulf says one of the longest moose migrations in the world goes right by his house.

I had met Ulf at the Savory Institute conferences in Boulder, Colorado, and London. Ulf is well over six feet, with light reddish hair and a broad smile. He is at once cheerful and somewhat cynical, paired qualities I attribute to his being an avid reader and his bent toward looking at things through the lens of deep history. He can see beneath what's presented (hence the cynicism) yet trusts nature's resilience and capacity to change over time (and so the good humor). The name *Ulf* comes from the Old Norse word for "wolf." *Ullring* derives from Ull (or Ullr), the god of skiing and archery in Norse mythology. In the pantheon of the far north, he says, "There are gods who love skiing and cold."

As soon as I'd received the speaking invitation, I reached out to people I knew through the Savory network to alert them I'd be in Norway in early May and said I hoped to see a bit of the country as opposed to just flying in and flying out. Within a few email exchanges, I had an itinerary: Ulf would meet me in Trondheim, take me through the fjord region, and then put me on the train from Lillehammer to Oslo. There I would connect with Helge Vittersø, who would host me at his family's apartment and show me around Oslo, where I would speak at the Norwegian Farmers Union. The night before my talk, Helge stayed up late cooking beef bourguignon—using meat from fellow Nordic Hub member Trond Ivar Qvale's holistically managed farm—and opened his home for attendees to join us for dinner.

Before Ulf and I get on the road, we walk through Trondheim, which is lively on a Sunday morning. He shows me the bronze statues of violinist Arve Tellefsen and Olympic speed-skating champion Hjalmar Andersen, sons of the city forever cast in their pursuits. Once we have a view of the outskirts, Ulf points toward the north and east, where his father grew up during the Nazi occupation. As a child Ulf heard stories about this fearful time—especially the food shortage, as German soldiers billeted in small towns commandeered local farmers' crops. "All Norwegians had a connection to a farm until recently," he says. (Reindeer also played a role in World War II: Their ability to pull weight across rough arctic territory proved pivotal in the Petsamo-Kirkenes Offensive in 1944, during which Russian troops drove German troops from northern Norway.[14])

Thanks to oil and other resource extraction, Norway ranks among the world's wealthiest countries. Ulf's story is a reminder that it wasn't always so. As we travel we see more evidence of the country's economic and ecological vicissitudes, past and present. On our way west to fjord country, we drive through steep hills covered with evergreens. To me it looks pretty, and I assume this is the natural landscape. Ulf corrects me. "A lot of this is Sitka spruce from western Alaska," he says. Favored for their tolerance to wind and salt, the trees were widely planted near the coast. "Sitka spruce is now blacklisted. It spreads and everything else is shaded out. The native plants die out."

Over the past century and a half, he says, spruce was planted on old grazing lands—areas where the grass fields were small and steep inclines made for difficult access, especially for cattle brought here for summer grazing. "Modern cows do not graze on rangelands as they once did. There is less open grazing land than in five thousand years, maybe always," he says. "Where there used to be cattle, there are now deer. They graze and browse—and herd—but not like cattle. People who raise cattle want to feed them. Grass is grown like crops, fertilized and sprayed to get as much as possible. Plus there's imported soya from Brazil and grain from the United States. Barley used to be the bread grain. Now it's used for feed. During World War II, Norway produced almost all of its food. Now it's closer to 40 percent."

In other words, as with pretty much all of the developed world, Norway's prosperity is sustained by soil fertility borrowed from elsewhere.

Farming here was never easy, Ulf says, particularly given the short growing season. He indicates the left side of the road where land for growing would be in nearly perpetual shade, due to the angle of the hills.

"Farmers from here likely went to the United States," he says. Starting in the mid-1800s, there were waves of immigration from Norway to North America. He tells me some members of his mother's family went to Iowa in the 1850s. Maybe they had been among those relegated to the dark side of the road and struggled to get a crop, even in a good year.

The vista opens up and we are among the fjords: waterways that draw cool, blue lines through landscapes of mossy green hills and fields. In the distance are steeper rocky peaks, the tallest ones topped with snow; in the foreground occasional red farmhouses, clustered together or alone. After coming around a curve, I note a sizable cove where several large circles sit on the water surface: net pens for fish farming.

"This is farmed Norwegian salmon, making some people rich," Ulf says. He makes it clear that he is not a fan: "Salmon used to be healthy, and got their orange-red color from eating small crustaceans. Fish farms are polluting, and they use artificial coloring and antibiotics." He adds that compared with wild salmon, the farmed fish has a less favorable balance of omega-3 and omega-6 fatty acids—higher omega-3 is what you want—and is therefore not as healthy.

While the scattered fjord settlements are picturesque, Ulf says that these communities are becoming abandoned "It's hard to keep the farms going, as they're spread out in the landscape," he says. He says that animal husbandry in Norway has traditionally spread in and out of the fjords, which were often used for winter pastures because here they don't have deep freezes or permanent snow.

In the past, people often traveled by boat through the fjords, which form from glaciers expanding and contracting over time, but roads have made that kind of travel more or less obsolete. There are cruises and train trips for sightseers, but for ordinary getting around, the ferries people had relied on have been replaced by tunnels, including some so lengthy you think you will never see the sky. "This has become a country of tunnels," Ulf sighs. "But this way you don't see the beautiful fjords." He has planned ahead so we catch a ferry, a twenty-minute ride during which we can relax and watch the water stream past.

For the thrill-seeking fjord traveler, nothing beats the Atlantic Ocean Road on the route to Kristiansund, with eight narrow bridges

spanning the sea and whose roller-coaster sensation has earned it
a prominent place on the website www.dangerousroads.org.[15] The
road twists and bends above the sea, briefly finding solid ground on
rock outcroppings before lifting again. "I've never been here, but I've
watched clips on YouTube," Ulf says as the road swerves and even
in the car we feel the whip of wind. "See how the water washes the
road? People come out this way just to do this."

Ulf meets me at the symposium at Kunsthall Trondheim. Due to
previous commitments I could only attend the second day. While I
missed in person the presentation by Marie Roué, an anthropologist
at the French National Centre for Scientific Research in Paris, I later
watched the videotape of her talk on Sámi ecology and the science
of snow. Roué, a French Canadian scholar, has done fieldwork in
Norway and Sweden with the Sámi and with Cree First Nations
and Inuit in Canada. She describes the complexity of winter reindeer
herding, which requires continually monitoring snow and the state of
the pastures, as well as understanding the nature of snow and how it
is experienced by the animals.[16]

Reindeer herding is an existential art, says Roué. What she means
is that a herder's knowledge is based on the essence of snow, which
is impermanence. The herders are always making decisions according
to evolving conditions, including that of the animals. "The reindeer
has to eat. For many months—about nine months, depending on
the year—he has to eat lichen and other things which are under the
snow," she says. And so the condition of the snow, wind, and sun
are important. The reindeer "has to dig, he has to smell. If there is
a crust of ice on top of the snow, he might not smell his food. And
if the crust of ice is very, very thick, he cannot dig." She adds that
modern scientific assessments of how much land is required for a
given number of reindeer are often predicated on an "average year,"
which might not reflect the reality at any given time.

Roué says scientific calculations miss other nuances about how
reindeer respond to ecological variations. In addition, reindeer are

now in areas exploited by industrial forestry. "When you exploit the forest, the composition is not the same. If you exploit the forest and you cut it so you can have pines, all the same, everywhere, [the snow] will not fall in the same way as when the snow is on the crown of the tree." Today's Sámi know how the impacts of industry ripple out into the environment, she says. She emphasizes just how vulnerable herders are to the unpredictability brought by climate change, "because the temperature is not constant. When it goes cold and warm again . . . then it melts, there is water and then there is ice when it is cold again." Conditions like this can destroy pastures.[17]

As I watched this, I couldn't help but think Roué's presentation would have been helpful for government scientists seeking to understand the ecological impact of reindeer herding, given that this was their stated concern.

———

Of course, the pressure on Sámi herders like Jovsset Ánte Sara is only partly the result of misconceptions about the effect of reindeer on the landscape. It is also about power: specifically the power differential between a wealthy Western democracy oriented to the global economy and an indigenous community that's small in number and yet occupies a large geographic area—one that happens to be rich in natural resources. In Trondheim, Ánde Somby, a Sámi and associate professor at the Faculty of Law at the University of Tromsø, took on this issue in a powerful lecture about what happens when a "prey culture" meets a "predator culture."[18]

Somby specializes in indigenous rights law at the University of Tromsø (also called the Arctic University of Norway), the northernmost university in the world. He also performs *joik*: a Sámi music tradition in which the vocals may invoke another person, an animal, or a feature of the natural environment. During a break, Somby approached me and thanked me for the talk I gave that morning. He told me he has observed that cloudberries do better when reindeer graze them.

He wore a striking brown wool hat, snug around the forehead and knit into four peaks. I learn that this is a Four Winds Hat, traditional

headwear for Sámi men. Usually, like Jovsset Ánte's outfit, it is dark
blue with red and blue trim. Somby's seemed like the professorial
tweed version, matching the subtle, dark plaid of his suit jacket. On
the other hand, his bright orange trousers hardly fit the look of the
staid academic. Clearly, this is someone who goes his own way.

And in his lecture where he goes is to allegory: specifically, Aesop's
fable of the wolf and the lamb. In the story, the wolf seeks an excuse
to devour the lamb, basically to show that the lamb had it coming.
The wolf claims the young sheep mucked up the water, then accuses
the lamb of insulting him—neither is true. The wolf could have easily
made a meal of the lamb without the excuses, so Somby asks: Why
did the wolf bother to establish that the lamb had wronged him?
His answer is that a tyrant will try to rationalize the harm they do to
those they prey upon.

Somby edges toward the Sámi's predicament. When a predator
and a prey culture meet, he says, the predator culture's goal is access to
natural resources that the prey culture controls. How does the predator culture achieve this? One approach is to convince those that are
in the way that they should share or give up their bounty. This is right
from the wolf's playbook. Somby describes the "family metaphor."
For instance: "In our National State we are all a family. And we all
need to contribute to our happy family." The implicit threat is: Don't
stand in the way of what's best for our cozy household. To Sámi
people living on the coast, the message is: "We have to think cost-
efficiently. You are there with all your small boats. It's so inefficient,
the way you fish. The big trawler—that's much more cost-effective.
Why not hand the right to fish to them? That's very good for our
happy family."

Another predator tactic, says Somby, is to shift language, like using
legalese: "In a legal sense, the indigenous people had no concept of
ownership so it's very good that the civilized people came in and
structured the situation." Or propaganda. "You call the mining . . . the
new thing, the new, beautiful thing that is coming," compared with
reindeer herding or subsistence fishing, which is "old, a little bit dirty,
[a] traditional thing [that] belongs to the past."

Case in point: The Norwegian government is urging its people to welcome the mining development and mineral extraction as the route to a prosperous future. In early 2019 Norway approved the construction of a copper mine described as "one of the most environmentally damaging projects in [the] country's history."[19] The site is in Finnmark, the country's largest county, on land replete with not only copper but precious metals and offshore oil and gas. It is also in Sápmi territory, on land where Jovsset Ánte's reindeer migrate and mate during the fall.

A further tactic of a predator culture, says Somby, is to render the prey culture invisible: between the lines of the national story. The point, he says, is that they are not in the text but "hidden away. Nobody sees them." He cites research showing that most students and teachers in Norway today are unaware that between the years of 1850 and 1980, Sámi children were sent to boarding schools where they would not hear their own language, under an official governmental policy of integration, or "Norwegianization."[20] Actions like this, he says, alienate prey cultures from their own backgrounds, undermining the knowledge and cultural heritage that defines who they are.[21]

Environmental anthropologist Hugo Reinert, a senior research fellow at the University of Oslo, argues that the government's misunderstanding of reindeer ecology and the disempowerment of Sámi people are deeply connected, if not one and the same. According to Reinert, the menace of "too many reindeer" has been a common refrain among Norwegian experts for a long time. He says the notion of controlling the reindeer is a stand-in for reining in the Sámi people that have thus far defied control. The argument in recent years, he says, has focused on damage to the iconic tundra landscape. To urban Norwegians in the south, Sápmi territory is lawless and chaotic. Since the Sámi are seen as incapable of regulating their herds, the government must take action.

"Expediently, the escalation of this failure narrative has coincided neatly with the escalating interest of national and international actors in 'developing' the tundra, 'realizing' its 'economic potential'—a potential that only becomes more attractive with the global depletion of easily available mineral resources, say, but which pastoral land claims stay in the way of," Reinert writes.[22]

The result, says Reinert, is a situation where policy makers assert there are too many reindeer, based on research that, he suggests, may be designed to reach this conclusion.[23] Upon hearing this message repeatedly, the general population believes "saving" the reindeer requires drastically culling their numbers or geographically containing them in a way that, in fact, damages the ecosystem. When the Sámi disagree, it becomes evidence of their primitive and unruly nature, leaving them doubly violated: The Sámi lose their reindeer and their livelihoods, and their identity, which rests on cultural knowledge of herding, honed through many generations, is dismissed and devalued.

For Reinert, the predicament of Sámi reindeer herders is exemplified in Máret Ánne's *Pile o Sápmi Supreme*: that shroud of metal and bone placed at a respectable remove from Norway's high court. How many people around the world regard their nation's seat of power from behind a metaphorical screen? He writes of the installation: "I had never seen anything like it; it tore open the asphyxiating mildness of national debates, manifesting in a torrent what the quiet, softspoken colonialism of the north—patient as it is, understated, polite, and bureaucratic—kept under wraps . . . The skulls manifest, physically, a siege that has gone on for centuries."[24]

Our second day of touring, on our way to Lillehammer, Ulf's home base, he brings me to the Norwegian Wild Reindeer Centre Pavilion, a feat of architecture on the outskirts of Dovrefjell National Park. The pavilion's curvy, wooden seating area is nestled within a cool, glass-and-steel exterior presiding over a stark, snowy hillside. Ulf explains that the pavilion is the work of Snøhetta, a Norwegian architecture design firm of international renown. Among their other projects are the Alexandria Museum in Egypt and the 9/11 Memorial Pavilion at the World Trade Center site. Snøhetta is also the name of a mountain in the distance, the highest in the Dovrefjell range that divides the eastern part of central Norway and the region around Trondheim. The Dovre mountains figure prominently in Norwegian history and folklore, including its rich mythology about trolls.

The Dovrefjell National Park is the best place to see wild reindeer not just in Norway but all of Europe, so we decide to walk down the hill from the pavilion. A lovely idea, but for the fact that there's a fierce wind and it's bitterly cold. I am conscious of every step. It is the second week of May and I'm holding my jacket hood tightly around my chin. As I trudge on, I hear the rippling sound of wind against fabric and the crunch of snow under my shoes.

"There are musk ox in these mountains," Ulf says casually. "They are very dangerous." These large, shaggy-haired mammals were denizens of the Zimovs' cherished Mammoth Steppe ecosystem and died out in Northern Europe after the Ice Age. They were reintroduced to the park in the late 1940s and early 1950s from Greenland and Baffin Island. If I spy a musk ox, I should give it plenty of space, Ulf warns.

We trek onward and Ulf points out some bushy plants, rather lonesome-looking things. He says these are dwarf willow and birch and juniper, explaining that trees stay small in this difficult climate. We do not see any reindeer (or musk ox, for that matter). I do, however, find a tuft of grayish fur and put it in my pocket. Ulf says this would be an area for ptarmigan or willow grouse hunting.

The cold saps our ambition (or at least it does mine), so rather than continue on our hike, we turn back and follow a short trail marked with historic plaques. The bones and tusks of now-extinct large mammals were found nearby, notes a sign describing the area in prehistoric times. The Vikings traded reindeer products like antlers and hides with the English, Ulf says, as we pass signage for the eighth to eleventh centuries. They learned to trap reindeer, and the animals' population nearly crashed. During the Black Death and other plagues in the mid-fourteenth century, Norway's human population crashed as well, by half.

———

Not surprisingly for an ecologist who lives in a land of reindeer and moose, and with reminders of their tusked and thick-furred cousins buried in sand and soil, Ulf has an abiding interest in megafauna. He is fascinated by the role large mammals play in how landscapes evolve over time, including Norway's heathlands and montane woodpastures. Over the course of our long drives, he shares with me what he's learned from reading and from the 2014 Oxford Megafauna Meeting, which he followed online.[25]

Up to some ten thousand to forty thousand years ago—"a blink of the eye in terms of earth's history"—the world was full of large animals. Elephants, rhinos, and hippos rambled around Europe while North America was home to mammoths, saber-toothed lions, and beavers more than twice the size of beavers that live today. Australia, in particular, enjoyed a menagerie of outsized wildlife, like the diprotodon, a marsupial that resembled a giant wombat and may have weighed as much as three tons. While the exact reasons for diprotodon's extinction between twenty-five and forty thousand years ago remain a mystery, most experts think it was likely due to a combination of hunting and climate change.

The Oxford Megafauna Meeting explored the environmental implications of the loss and the possibility of reintroducing wild species or their analogues to reinvigorate landscapes. Attendees and presenters shared the view that appreciating the role of megafauna in past environments is essential for interpreting current animal-land-climate dynamics. As Yadvinder Malhi of Oxford's School of Geography and the Environment writes with his colleagues, "Much of our current understanding of ecosystem ecology and biogeochemistry has been developed in a world artificially depleted of giants."[26]

After I returned home to Vermont, I read up on the conference. I tried to stay mindful of the big questions of the ecological role of megafauna while also letting my mind wander into a "Where the Wild Things Are" realm of gigantic beasts moving through forests and plains. The thought of megafauna past and present does capture the imagination and awaken the childlike sense of wonder at the bigness around us. How many children's books feature talking elephants and friendly bears, in part as a means of taming what is big and wild and scary? It seems we are wired to be fascinated by megafauna. Indeed, our species evolved alongside creatures far larger than us.

I was also curious, however, to read some of the scientific nitty-gritty. One paper I read that stemmed from the conference articulated several means by which large animals influence the environment. Megafauna are loosely defined as mammals whose weight exceeds one hundred pounds (some use one hundred kilograms as the marker). For a land mammal to qualify as a *megaherbivore*, it would have to weigh more than a thousand kilograms, or twenty-two hundred pounds.

Being big can be a strategy to protect against predators. In order to stay big, megaherbivores must eat a lot, and in eating a lot they alter their environment. Therefore, they are "ecosystem engineers" through their very existence. "In trophic terms, megaherbivore populations are generally considered to be limited from the 'bottom-up' by food availability . . . and thereby exert strong 'top-down' control on vegetation structure and composition."[27]

In eating, trampling, breaking, and inadvertently damaging plants, large herbivores help determine the balance of woody versus herbaceous vegetation, or trees and shrubs as opposed to grasses and forbs. We've seen this is the case with reindeer: Their shrub-browsing and trampling ensures the predominance of lighter-toned grasses and lichens that reflect sunlight and keep the landscape cool. In North America and southern Africa, large herbivores—buffalo in North America; wildebeest, kudu, giraffe, and whatnot in Africa—maintain open savanna. Over time, their activity built the rich, fertile soils, the *mollisols*, of these regions.

A marquee megaherbivore like an elephant can fuel an entire ecosystem. For an elephant, it's nothing to knock down limbs from, say, acacia trees. In doing so, it shakes out edible pods and leaves to feed other animals. Simply by walking, they create pathways, and by digging for water, they create waterholes. Then there's dung. A good-sized pachyderm will generate fifty kilograms (110 pounds) of it a day.[28] Elephant dung provides a food source for birds, insects, small mammals, and primates, as well as a habitat for a variety of insects and even amphibians. The dung is full of seeds that get dispersed, often across significant distances.

In other words, megafauna are pivotal to nutrient cycling. "Nutrients that would be locked for years in leaves and stems are liberated for use through animal consumption, digestion, defecation and urination."[29]

This is particularly significant, the authors say, in areas with nutrient-poor soil or dry or cold places "where megafaunal guts can act as giant warm and moist incubating vats that accelerate otherwise slow nutrient cycling." With their large appetites, copious waste, and wide-ranging travels, large-sized creatures distribute nutrients through the landscape, seeding smaller pockets of fecundity and biodiversity.

Another mega-mammal that drives nutrient flow is the whale. Author and University of Vermont researcher Joe Roman has explored many ways that whales feed and provide habitat for life in the sea. For example, great whales—species like baleen and sperm whales—graze at depths and then release their waste, delicately termed "fecal plumes," near the ocean's surface. This spurs the growth of plankton, the basis of the ocean food chain.[30] This series of trophic events has been described as a "whale nutrient pump." And, incidentally, this ocean plankton consumes carbon and releases oxygen, producing much of what we breathe.

When a whale dies and falls to the ocean floor, it banks a lot of carbon, and becomes food for many other organisms. The carcass will house communities of invertebrates, such as clams, lobsters, and crabs. "Dozens, possibly hundreds, of species depend on these whale falls in the deep sea," says Roman. Like elephants, whales carry their nutrient bounty a long way, especially since they generally travel far to breed.

Land-based megafauna help determine the impact of fire. "In many ecosystems, megaherbivores act alongside, or in competition with, an abiotic 'herbivore,' fire."[31] In other words: Plant-eating animals and fire both consume vegetation, and thus serve a similar ecological function. As Allan Savory has pointed out, when an animal consumes plant material, it is processed biologically and the carbon remains in the ecosystem. By contrast, when a fire consumes plant material, it is processed chemically and the carbon is released into the atmosphere. Even aside from fire's destructive aspects, recycling organic matter through the food chain is preferable from an ecological standpoint.

While there is an ecological role for fire, frequent or large-scale fires promote fire-dependent plants and therefore perpetuate the conditions that create it. This tips the landscape toward aridification and makes fires more difficult to manage. This has happened in many parts of the world, including in the western United States, Southern Europe, and Australia.

In March 2017 I visited a very green Northern California after an unusually rainy winter. While some locals were enjoying the relative lushness and relief from years of drought, others expressed concern that all the vegetation meant a lot of fire fodder. It turned out to be a devastating fire season, and the following two years were even worse. As Malhi and his colleagues write, "In drier systems, or where human activity has greatly increased fire ignition frequency, the loss of grazers can increase grass fuel loads and lead to a shift to a fire-dominated ecosystem."[32] For this reason, managing fire-prone landscapes with grazing animals is extremely important.

A species' environmental impact is never simple or linear, and this can lead to faulty beliefs. Beavers, for instance, have long been considered pests because they take down trees and build dams where people don't want them. Only recently have conservation professionals started to acknowledge their role in creating and maintaining healthy wetlands and waterways.

Ideas about so-called nuisance species can become entrenched. This is a challenge for Chris Henggeler, an Australian farmer who manages Kachana Station in the rugged Kimberley region, and considers the wild donkeys that roam the terrain to be allies against unwanted fire.[33] His hope is to eventually incorporate them in a multispecies herd to heal and to manage larger sections of land. The Western Australia government, however, is determined to eradicate them.

Used as pack and draft animals, donkeys were instrumental in the northern settlement of Western Australia in the nineteenth century. Once motorized transport was introduced, the donkeys were no longer needed and many were set free. Donkeys are hardy, clever, and resourceful creatures, and they live a long time, often upward of forty years. They thrived, formed large territorial family groups, and became a feature of the landscape. As their numbers grew, rural state governments instituted eradication programs.

Over the last three and a half decades, the state of Western Australia has killed more than five hundred thousand wild donkeys in its northern reaches.[34] The donkeys are often culled with what's known as the Judas technique, in which radio transmitter collars are fitted to selected jennies (females). Because donkeys are intelligent and social creatures, the jenny will seek out a group of other wild donkeys to live among. Her new friends are then shot from a helicopter while the "Judas" donkey is left alone, so that she will then lead the cullers to a new group, and begin the process again, in a terribly cruel, if effective, means of control.[35] The government regards the donkey as a problem animal, with the potential to overgraze, foul water sources, introduce weeds (via seeds carried in their dung and on their tails), and compete with livestock when food is scarce. They have been known to damage fences and water infrastructure on farms.

At Kachana Station, Chris saw something different. Kachana is 77,500 hectares (almost two hundred thousand acres)—larger than Singapore—and many sections are remote and difficult to access. The mission of his Kachana Pastoral Company is to holistically manage the land in order to restore it and minimize fire damage. With limited staff and livestock (in Chris's parlance, upper and middle management, respectively), there was a limit to what they could accomplish. And certain areas, like upper river catchments, were ill suited to cattle. Assessing these constraints and thinking about how they could work within them to intensify impact, Chris and his late father, Bob, realized that donkeys could at least be part of their fire mitigation strategy.

"The Donkeys are not yet run intensively with the cattle herd, but they do a great job maintaining low-fuel zones around our model areas and assist us in wild-fire mitigation in the areas where we cannot justify the use of cattle. We observe a whole of lot other ecological benefits associated with their behavior including the establishment of perennial ground-cover in areas that have been bare for at least three decades," Chris writes.[36] He adds that cattle and donkeys' intense grazing "makes

life easier for the 'nibblers' (kangaroos, wallabies, and smaller marsupi-
als) to follow up with keeping fuel-loads in check."[37] This is important,
as marsupial populations in many parts of the region are in decline.

The rush to destroy wild donkeys has led to an increase in senes-
cent grasses, Chris believes, and he says this, together with a run of
relatively high-rainfall seasons since 1994, has likely been a factor
in repetitive destructive wildfires. He is putting together a research
program, the Wild Donkey Project, which includes soil sampling
and other land monitoring. "We're inviting scientists to study these
donkeys and what we're finding is that all wild animals have a
function out there in the landscape and we can actually influence
those functions and enhance the health of ecosystems," he told the
Australian Broadcasting Company in 2018.[38] They are using a low-
stress "pressure-release" approach to train the donkeys to cooperate
and move to where they are needed.

Chris's commitment to protecting Kachana's donkey population
has led to conflict with the local and state governments. In May 2018
the Kimberley Rangelands Biosecurity Association wrote to Chris,
"Our firm conclusion is that we are unable to support the use of
a declared species, in particular donkeys, in any way as a landscape
management tool."[39] Chris has worked to educate the government
and the public about the ecological benefits of managed wild donkeys.
He has taken out ads in regional newspapers and invited other land
managers into the debate. The plight of the donkeys has attracted
interest from international broadcast and film media. The effort has
postponed the cull. As of this writing Chris reports that Kachana and
the state officials are at a "stand-off."

Scientist Arian Wallach's perspective on what Chris Henggeler
calls Australia's "new megafauna" also differs from the standard line.
Wallach, an ecologist originally from Israel who teaches at University
of Technology Sydney (UTS), studies the impact of animals on the
vitality and biodiversity of ecosystems. She has focused on the dingo,
a wild canine related to dogs that, like the wild donkey, is maligned

and subject to population control—meaning that it is routinely shot, baited, or poisoned.

Wallach's research suggests that that apex predators like the dingo are important to ecosystem health. She founded the Dingo for Biodiversity Project to promote predator-friendly farming and a broader understanding of the role dingoes play in the Australian landscape.[40] The program is part of the Centre for Compassionate Conservation, which promotes respectful and humane treatment of all wildlife, housed at UTS.[41]

Wallach's work challenges assumptions about how animals interact with their environment. It also raises questions about the foundational ethics of conservation as a discipline. She contends that animals like the wild donkey, whose geographic presence transcends the boundaries we've set for them, are "invisible megafauna." In a talk titled "Feminist Ferals," presented to the New Zealand–based online conference Feral, she discusses what this means in terms of our relationship with the natural world, and for the wild donkeys themselves.[42]

For one, she poses the question: *Who decides which animals are native to a given landscape?* Conservation often seems to have selected as the natural ideal the way a place appeared when colonial explorers first arrived, what she calls the "white dude moment," quoting author Emma Marris. Given that indigenous people had been managing many landscapes for millennia, the arrival of explorers or colonists is an arbitrary distinction. And yet this fixed point has often determined what's considered native. As for animals like wild donkeys, which are not native to Australia but have found there a viable niche, she says, "They mess up our idea of nature as a Garden of Eden."

The donkeys are also, she says, missing from scientists' and governments' assessment of current ecology. Since they don't "belong" in Australia, their impacts either are not considered or are presumed to be adverse. For instance, the Australian government points to wild donkeys' dispersal of seeds as a negative impact even though it enhances biodiversity.[43] The relative success of pioneering species like musk ox in Norway and wild donkeys, camels, and water buffalo in Australia, she says, offers important information about natural systems and can therefore inform policy. "The very populations excluded from conservation are arguably manifestations of a key mechanism that enables life to adapt to change. The stories of flourishing displaced wildlife shatter the idea that everything touched by humans is forever bound to us and humbles our view of ourselves as so-called wildlife managers."[44]

With animals like wild donkeys in our collective blind spot, we are less prone to intervene when they are treated inhumanely. Wallach points out that donkey skin gelatin is a key ingredient in *ejiao*, a Chinese medicine. To accommodate the large market for ejiao, she says, about two million wild donkeys are killed every year, mostly in Africa and Asia. Many are penned up and suffer cruelty and neglect. This receives scant attention, however, because donkeys currently fall between the cracks of what society values: They are seen neither as productive working animals nor a legitimate part of nature worthy of conservation.

Wallach says that although the Australian government and conservation science see wild donkeys as an invasive species, one can regard the animals' five-million-strong presence there as a successful "rewilding event." And that the advent of such mammals in new environments has an ecological significance that, due to their erstwhile invisibility, has been overlooked. "Introduced megafauna are a wonder of the Anthropocene hidden in plain sight," she writes, adding that inadvertent rewilding has maintained populations of animals that might otherwise have gone extinct in their presumed habitats. For instance, the dromedary camels that live in Australia's deserts are the world's only existing wild herds.

These "invisible megafauna" are likely filling ecological niches that have been left vacant since the dramatic loss of species in the late Pleistocene, she tells the Australian Broadcasting Corporation, adding, "Introduced megafauna are probably both replacing old and adding new ecological functions."[45]

Rather than a nuisance, can the wild donkeys' presence be filed under biodiversity? Wallach says that when we look through this lens, "we also discover that the world is in some respects wilder today than it has been for thousands of years . . . Introduced and feral megafauna have numerically replaced between 15 percent and 67 percent of species lost in the Pleistocene." Australia lost all of its megafauna tens of thousands of years ago, she says. Today there are eight species, including wild donkeys.

The vision of Compassionate Conservation resonates with me: to create societies that include and honor all species—to appreciate the intrinsic value of all life-forms and to respect their sentience and agency. At a time when much of nature is under strain, it is painful to acknowledge the loss of animals, birds, and other organisms. There is additional astringent bite when those beings are destroyed due to peremptory decisions about which animals are worthy of saving, and which are not. This kind of routine culling is far more common than most of us realize. Yet it remains invisible because most of us trust the experts to do the right thing. People like Chris Henggeler and Arian Wallach in Australia and Máret Ánne and Jovsset Ánte Sara in Norway are bringing the reality of these policies into view. In doing so, they encourage us to look more closely, creatively, and openheartedly at our relationships with animals.

"We can simultaneously mourn for what the world has lost and cherish what remains and what is thriving," says Wallach. She says that indigenous peoples' views toward invisible megafauna communities tend to be more generous. She quotes an indigenous man who says of wild donkeys: "Yah, but they belong to this land now. We can't push them out, any of them: camel, donkey, kangaroo, emu. They belong to this land."

Wallach's critique of traditional conservation is that it denies animals agency and subjectivity. Sometimes, when I'm immersed in academic papers, I pause to remind myself that these are real beings and not just statistics or behaviors. Each is magnificent in its own way: the fleet reindeer moving in vast numbers across the snow; the gregarious and clever wild donkeys with their soulful eyes; the handsome and stealthy dingo with its frolicsome pups. These animals are worthy not just as representatives of a species but also as individuals, for their "is-ness": the beautiful fulfillment of what and who they are. As Wallach says, the dingoes do a better job of maintaining the populations of local wildlife with their predation than humans can ever do with our radio tagging and air shooting.

That is to say, we can never approach the wisdom of these animals.

In my travels with Ulf, he also described to me the work of Frans Vera, a Dutch biologist who claims that rather than closed-canopy forest, primeval lowland Europe was mostly savanna. He says that large herbivores like wild horses (tarpan) and cattle (aurochs) helped sustain a woodland-pasture system that was productive and diverse and allowed for sun-loving trees like oak and hazel to thrive. This challenges a long-held notion that in Europe ecological succession inevitably leads to dense forest.

Vera is known for his rewilding experiment at Oostvaardersplassen, dubbed the "Dutch Serengeti," in the Netherlands. In the 1980s he began bringing cattle, horse, and deer that best approximated their extinct, wild counterparts to a five-thousand-hectare (twelve-thousand-acre) tract of land east of Amsterdam. In many ways it has been a success. Birds, including rare raptors and marsh species, showed up, as did foxes and muskrats. Oostvaardersplassen became an attractive wetland preserve.

However, the hands-off management of herbivores led to dramatic population fluctuations. With mild winters, animal numbers grew. Harsh weather meant starvation and death. In 2018 there was a public outcry. People protested and brought in bales of hay, while Dutch state foresters culled animals so that they wouldn't suffer.[46] One complication is that Oostvaardersplassen was only partially wild: The site lacked large predators to manage the population of grazers. Currently beset by lawsuits, Oostvaardersplassen's fate is unclear.

Still, Vera's insights remain compelling and the reserve, which demonstrates interdependence among herbivores, trees, and wetland birds like geese, provides useful lessons. British author Isabella Tree was inspired by Vera's work, and undertook a similar project at her Knepp Estate in West Sussex, which she describes in her book, *Wilding*. In 1987 Tree's husband, Charles Burrell, inherited fourteen

hundred hectares (thirty-five hundred acres) of arable land, parkland, and historical buildings. The couple moved there to raise their
children and set up a conventional grain and dairy farm. By 2000
they realized that it made no economic sense to continue farming
conventionally, and they let the operation go. Two years later the
property was returning to prairie conditions. She writes, "The land,
released from its cycle of drudgery, seemed to be breathing a sigh of
relief. And as the land relaxed, so did we."[47]

Tree and Burrell visited Oostvaardersplassen, invited Vera to
observe their land, and decided to embark on a similar venture. They
introduced several animal species, including Tamworth pig (proxy for
wild boar), English longhorn cattle, Exmoor ponies (beautiful, with
floppy bangs), and red, roe, and fallow deer. Between the presence
of the animals and the absence of chemicals used to control livestock parasites, insect and bird life began to rebound: nightingales,
turtledoves, skylarks, willow warblers; multiple species of bats; and
charismatic butterflies like purple emperors and fritillaries.

The revitalization of the land and the emergence of more bird,
insect, and plant species continues to amaze Isabella Tree. She opens
a 2018 5X15 talk with the soft cooing of turtledoves, noting that
Knepp is the only place in England where their numbers are rising.[48]
She says bringing grazing (and rootling) animals onto the property
was like "injecting dynamism into a flat landscape. Suddenly there's
energy there again, and it's like putting a glider up into the air.
Extraordinary things start to happen. Things start to fly."

One impediment to more widespread rewilding, she says, is the
favored aesthetic in which everything, even nature, should be tidy, a
preference that arises from people's desire for control. We've grown
up with this "Victorian corseted culture," she says, adding, "We have
to sit back and let nature be in the driver's seat." This includes learning to live with nature's boom and bust cycles. She calls rewilding
"restoration by letting go." It is an exercise in learning to appreciate
the hidden potential of our landscapes, she says, as well as an opportunity to "rewild ourselves."[49]

FIVE

From Scarcity to Abundance

Reclaiming Hawai'i from Its Plantation Past

> *In our philosophy, the land and the people are one.*
>
> —Kealoha Pisciotta,
> of Mauna Kea Anaina Hou[1]

The Alexander & Baldwin Sugar Museum in Maui is housed in a former plantation supervisor's residence across the road from a decommissioned sugar mill. It is a nice, sharply sunny day when I visit, so I linger outside. Strewn about the grounds are pieces of equipment, relics of sugar's heyday: a huge metal cog or "bull gear"; a cane hauler that can tow sixty-four tons; and "cane grabs" that look like gigantic claws from a B-grade horror film.

Inside, the museum has a cheerful air, offering "fun facts" and presenting "a living legacy" of plantation life. But it is hard to sweet-coat the trenchers, treads, and crushers and the harsh labor—the mechanical and social violence—that bring the honeyed white crystals to our teaspoons.

One exhibit recounts the tale of Maui, the trickster god who is the island's namesake. According to legend, Maui lamented that the sun sailed too swiftly over the island for its people to get anything done. He climbed to the top of the Haleakalā (House of the Sun) volcano and lassoed the sun so that it would stay longer in the sky. The exhibit plaque explains that this is why the island has enough sun to grow sugarcane. Of course, a trickster's tricks are never straightforward.

There's always some mischief—of the "be careful what you wish for" variety—involved. Maui's ample sunlight has indeed created abundance. But rather than serving those who dwell there, much of that bounty has left the island to enrich others, leaving trouble in its wake.

When you think of a Hawaiian island, what comes to mind are sparkling beaches, graceful palms, strands of fragrant flowers strung together as leis. The fantasy is sun-washed, touched by ocean breezes, possibly spanned by a rainbow, or *ānuenue*. For visitors, particularly those escaping harsh northern winters, the Aloha State is sublime. For those who live here, it is a bit more complicated. With its warmth and dependable moisture, plants like it here and grow really well. This has been both blessing and curse, the two-pronged gift of a trickster demigod.

The history of Hawai'i is generally divided into two periods: pre-contact and post-contact. The "contact" is the arrival in 1778 of James Cook to what he called the Sandwich Islands, an event that led to political turmoil and an influx of missionaries, merchants, and adventurers. These newcomers brought to the islands many new effects, including measles, tuberculosis, influenza, and venereal disease, decimating the local population. They also brought numerous alien species, from pigs, goats, and cattle to seemingly benign decorative plants, many of which devastated native flora and fauna.[2] Hawai'i's endemic wildlife was highly vulnerable. Local species had developed in isolation, thousands of miles from the nearest landmass, and so never evolved protective defenses, such as venom (for a reptile) or thorns (for a plant). There was no need. Biologically speaking, pre-contact Hawai'i was a realm of innocence.

Another import for which local people had no resistance was the concept of land ownership. Until the *Great Mahale*, or division (in other words: sale), of lands in 1848, there was *ahupua'a*, a system of land sharing in which wedge-shape sections ran from the mountains to the sea.[3] There was wisdom to this, as the segments represented functional watersheds and included forests, food-growing areas, and the coastline, and so were self-sustaining economic units. Like pre-1952 Saudi Arabia, there was a system that worked for the people and the land. The advent of private land in Hawai'i set the stage for Samuel Alexander and Henry Baldwin, two sons of missionaries, to purchase five hectares (twelve acres) in Maui's uplands in 1869 for $110, launching what became a sugar empire.

In Hawai'i and elsewhere, sugar cultivation has been ecologically disastrous. Commodity sugar production invariably involves toxic pesticides, fertilizer runoff, and diversions of natural waterways. The canes are burned to remove the outer leaves. This causes breathing problems and sooty skies—and, in Maui, "Hawaiian Snow," as people call falling ash. The industry also played a role in suppressing the Hawaiian language and the art of hula dancing, as well as keeping ethnic groups separate to discourage labor unrest. With the indigenous Hawaiian population so depleted from imported diseases, sugar companies recruited workers from China, Japan, Portugal, and the Philippines to work in the fields and factories. Hawai'i's diversity is largely a reflection of Big Sugar's considerable workforce demands.

In January 2016 Alexander & Baldwin (A&B), the state's last existing sugar company, announced that it would cease production on its site in Maui by the end of the year. This was huge news, signaling an end to the plantation era that for 150 years had dominated the island's landscape, economy, and culture. Activists and other engaged citizens were overjoyed. Here was a chance to reclaim the landscape and create a vision for the island's future. Jenny Pell, a permaculture designer and teacher living in Maui, says people can now start "breaking the pattern of plantation overlords," which, she says, has brought "nothing but degradation and pollution. It's been like this for 130 to 160 years." With the sugar mill closed and many large-scale pineapple producers decamped to the Philippines, Maui's farming landscape, she says, is "like a blank slate."

For my friend Charlotte (Char) O'Brien, this represents an opportunity to shift to regenerative agriculture informed by local Hawaiian knowledge and practices. Char is an agronomist who has raised dairy cattle for breeding stock in Wisconsin and is the founder of Carbon

Drawdown Solutions, Inc., which uses biochar to build stable soil carbon from various organic waste. Biochar is produced through pyrolysis, a controlled low-oxygen burn, and has multiple applications, including as a soil amendment, water filtration, and neutralizing odors. The heat generated in its production can be captured and used for renewable energy. With its ability to revive inert or damaged soil and essentially run the carbon cycle in reverse, biochar is seen by many as an important strategy for addressing climate change.

Charlotte also spent several years in Vietnam working with bamboo as a building material, vehicle for eco-restoration, and biochar feedstock. She lived in a little house on factory grounds along with other workers in a village fifty-six kilometers (thirty-five miles) from Saigon. She came to admire the culture in which, she says, families take care of one another: "It's not even a question. You don't find that much in Western culture." I met Char a few years back at a Biodiversity for a Livable Climate conference.[4] On Maui she has three grown daughters and three grandchildren, which led her to move here fifteen years ago.

Motivated by the land newly released from sugar production, Char and three friends formed Āina First, an association of farmers, scientists, and consultants, to respond to both the opportunity and the threat it presented. Many residents feared that GMO crops would fill the vacuum created by A&B's exit, and the members of Āina First wanted to ensure this didn't happen. The word *āina* means land, or "that which feeds us."

The initial plan was for Āina First to raise funds and purchase the land outright for ecological farming. However, investors also caught wind of the opportunity. In early 2018 seventeen thousand hectares (forty-one thousand acres) of A&B's former sugarcane land were sold to an operation called Mahi Pono (*mahi* means "to cultivate"; *pono* indicates right and just). It seemed to come out of nowhere. People asked: *Who is behind Mahi Pono?* Journalist Deborah Caulfield Rybak wryly described the firm as the "corporate love child of a union between Canada's Public Sector Pension Investment Board and a California-based agriculture company called Pomona Farming."[5] Rybak, a former *Los Angeles Times* reporter, traced Pomona Farming to Trinitas Farming LLC, a firm based in the Central Valley of California that has been linked with questionable water-use practices and beset by lawsuits, especially involving California's almond industry.[6]

Over the next several months, local farming and environmental groups paid close attention as details about the mysterious Mahi Pono venture trickled out. Observers sought to read meaning into new hires, which included people known in the community, and specifics in the sales agreement. There were reasons for ecologically minded citizens to be wary. For instance, Mahi Pono quickly got busy in the statehouse lobbying for water access. There were also reasons for cautious optimism. The language they used hit the right notes. And Larry Nixon, the new general manager, told Rybak he wanted more native trees and native grasses to regenerate the soil. He said: "We need butterflies, ladybugs . . . When the day comes that you have to scrape bugs off your windshield when you get home, then I'll be doing my job."[7]

Upon the sale's announcement, the members of Āina First were disheartened that the sugar property was spoken for. Char and I continued to correspond as Āina First's options shifted. She explained how local activism often entails protest and litigation, which is costly and sows dissension even when successful. Education and collaboration are often a better strategy. Regardless of whether Mahi Pono turned out to be friend or foe, she said, it was clear that citizens were mobilized and wanted a say in creating Maui's food and farming future—that environmentalists and indigenous Hawaiians were coming together in a newly energized way to restore the sugarlands and beyond.

———

In the late summer of 2018, things got real for people on Maui. The hurricane season was extremely destructive, despite the fact that no hurricane even made landfall. This echoed a warning from Robert Ballard of the Central Pacific Hurricane Center: "Just because a hurricane doesn't make a direct hit, doesn't mean it can't cause major damage."[8] Though downgraded to tropical storms by the time they brushed the island, Hurricanes Lane and Olivia brought fierce winds, drenching rains and subsequent flooding, downed power lines, and, in the aftermath, brushfires that burned several homes and left dozens homeless.

For local people, particularly sobering were the ten to fourteen days of empty grocery shelves. "Barges were circling around and then diverted," permaculture designer Adam Hicks told me. "If it had hit the port, it would have been a disaster." Adam, who spent many years as an IT professional in New York and Chicago, is the property manager where my husband, Tony, and I stay: a former pineapple plantation in Makawao where Adam is developing agro-forestry and other regenerative agriculture programs, including, possibly, a retreat center. He resides in a barn some distance from the house, sharing his dwelling and acreage with four cats, including a young male and female, Lane and Olivia, who were born at the time of the storms.

Our trip coincides with a Regenerative Agriculture and Gardening Workshop at the Hale Akua farm and retreat center, which has brought several Iowa agronomists to the island, as well as Steven Apfelbaum, founder and chairman of Applied Ecological Services in Wisconsin. Steven is an ecological consultant and author who, over his forty-year career, has designed and managed thousands of eco-restoration projects. Āina First engaged him as an adviser on restoring the sugarlands, which he regards as a "one-hundred-year opportunity." Our first evening in Maui, Āina First is sponsoring an event, featuring Steven, about food security in the age of climate change, for which three hundred people have registered. And so, after a ridiculously long travel day, Tony and I grab an off-schedule dinner at Da Kitchen in Kahului and make our way to the Maui Beach Hotel.

A tall Hawaiian woman in bare feet and wearing a lei strides up before the audience and offers a *pule*: an invocation to mark an occasion. She appeals to "an expression, an entity" . . .

... that is navigating all that is, including this moment: including this

> beautiful collective of the rainbow, of the hearts, that love *Maui nui a Kama*, that love the earth. And so we ask permission of our mother to come and hold us in the nectar of forgiveness. To shake, rattle and roll the quickening of the release of the trauma within our expectations.

It is at once performance and response, part poem and part chant. She pauses at certain junctures: words left suspended until, seconds later, she alights on a new thought. I feel stripped of my outsider status, as if the prayer has pulled me into this gathering of hearts—I belong here due to the simple fact of my loving the earth. The speaker now shifts to Hawaiian phrases and I notice her movements: hand swaying that evokes a wheel, a wave, a waterfall, the gestures of hula. She continues:

> We bless the intention of the speakers. We bless the intention of all that is collected here, above and below . . . We love her, we love Maui. And it is with that love, that breath of spirit, we are born into the refreshment of the rejuvenation of the generations to come. And so we cup our hands. We say *mahalo*. . . And we offer our breath in dedication to the essence of aloha.[9]

"Maui, what's your future?" Steven Apfelbaum asks, when he takes the microphone. He explains that we manage the carbon cycle by working with soil and plants, and that ecological farming presents the chance to do this right. He says restoration "is about jump-starting the tendencies of ecosystems to move into the future."

Steven notes particular concerns in the sugarlands, such as steep slopes that present an erosion risk and the possibility of saltwater intrusion. (To prevent the latter: Soil rich in organic matter ensures effective water infiltration.) It is important to determine where to do agriculture and where not to, he says. Not all land is appropriate for cultivation: Preserving land for wildlife, such as habitat for birds, makes agricultural land more productive. While taking land out of production in order to improve production may seem counterintuitive, it supports pollinators, creates a buffer for water and wind, and improves water flow.

Permaculture designer Jenny Pell shares the stage with Steven, as his local counterpart, and remarks that 90 percent of Maui's food is brought in from off the island, primarily via diesel-fueled barges. "We are so vulnerable," she says. "Hawaiian traditional farming and indigenous agriculture offer the best buffering for climate change. We can borrow from different systems." What's important, she says, is to "shift from monoculture production for export to polyculture production for local use." She mentions a project to convert the idle Pu'unene sugar mill, that hulking industrial relic I saw near the sugar museum, to a worker-owned cooperative: a food hub with restaurants, a farmers market, a commercial kitchen, and a makers' space called The Kuleana Co-op. This would address another local challenge related to food security, the lack of food processing capacity and support for food entrepreneurs. The word *Kuleana* refers to responsibility that brings you a sense of pride—a duty you embrace.

"We have to get back to traditional crops," she says. She notes *kalo*, or taro root, a prized native plant central to Hawaiian cuisine, and *ulu*, or breadfruit, a tree fruit that's nutritious and versatile and has been a staple among Pacific Islanders for thousands of years. She says there is evidence that the trade winds are failing, which means less and less predictable rain. This adds urgency to food security and restoration efforts.

Chip Fletcher, a coastal geologist at the University of Hawai'i's School of Ocean and Earth Science and Technology, has been warning the public about changes the state is likely to see in coming years. He has called attention to the waning of trade winds, noting that buildings in Hawai'i have been designed to take advantage of the winds and that without them, dwellings may become uncomfortably, even dangerously, hot. Failing trade winds are often associated with strong El Niño years, and such years are expected to become more frequent, says Fletcher. Other characteristics of El Niño years in Hawai'i, he writes, include drought, bouts of extreme rain, heat waves, active hurricane seasons, high ocean temperatures with coral bleaching, and more powerful winter waves that can cause flooding and damage beaches, roads, and seawalls.[10] People in the audience have no doubt observed many of these trends, even before the double whammy of Lane and Olivia.

Jenny closes by saying that long ago, back when the south of the island was forested, a cloud bridge connected the slope of Haleakalā with the smallest of Hawai'i's main islands, Kaho'olawe. Now desolate and dry, this eighteen-by-ten-kilometer (eleven-by-six-mile) island has historically been used for housing prisoners, grazing goats and cattle, and military target practice and bomb testing. She invites the people of Maui to dream of bringing back the cloud bridge, of connecting the islands via the moist *Naulu* winds, of recognizing what once was there and can be again.

Hovering over every discussion that day is the undetermined agenda of Mahi Pono and its enigmatic farm manager, Larry Nixon. Will he or won't he champion regenerative practices? Will he or won't he honor the promises he has made? Nixon has gone out on field visits with Steve Apfelbaum, and local activists including Jenny Pell, Susan Campbell of Āina First, and Mark Sheehan of Maui Tomorrow have met with the company team. Jenny reports that Mahi Pono has committed to cleaning up all watersheds on their properties by the end of 2020. Alika Atay, a former council member and representative of Hawaiian Indigenous Natural Agriculture (HINA), says Nixon told him he planned to put eight thousand hectares (twenty thousand acres) into regenerative farming, and wants the results of all soil tests made public.

One big question for locals is water. Alexander & Baldwin has been diverting water from East Maui streams since the late 1870s, before the 1893 overthrow of the sovereign Kingdom of Hawai'i.[11] One consequence has been lack of water for traditional taro farmers, many of whom had to abandon their fields as the land went dry. In 2016 taro farmer Ed Wendt prevailed in having stream water restored to his land, but only after decades of lawsuits.[12] Trust in commercial entities doesn't come easy in Maui. People here have been betrayed often enough before.

In 2014 Maui County voted in favor of a moratorium on GMOs unless and until they are scientifically proven to be safe. This despite Monsanto and Dow spending nearly $8 million to convince voters otherwise: an act of largesse that amounted to more than $300 per every vote cast against the measure. The people's victory was short-lived, however, as the companies sued. A federal judge ruled that the ban could not stand, determining that it was preempted by state and federal law. Anti-GMO initiatives on the islands of Kauai and Hawai'i (the Big Island) met a similar fate.[13]

Chemical agriculture has hit Hawai'i hard. The sunny islands, it turns out, are an ideal place to test GMO crops. While a company may be able to complete one crop trial per year in, say, Iowa, Hawai'i's beach-friendly weather means prime growing conditions year-round, which allows for three plantings a year. Most corn grown in the United States today, the bulk of which is used for ethanol or cattle feed, is genetically modified and comes from stock developed in Hawai'i. According to *Scientific American*, biotech companies "have turned the Islands into a sprawling living nursery for GM corn seed."[14] Sure, as the companies would claim, this has brought local jobs. But at what cost?

Bob Streit of Central Iowa Agronomy and his wife, Carol, are also staying at the Makawao house, so we sometimes chat during breakfast. Bob has been coming to Maui for years, as he's been researching and speaking about the consequences of agricultural inputs on soil and human health. He has lots to say, much of which is rough to hear. For example, he says that the actions of Dow, DuPont, and Syngenta in Hawai'i amount to a genocide.

Bob gives the example of Waiamea, on the island of Kauai, a picturesque area of green, rolling pastures that has been dubbed "poison alley." In 2013 and 2014 children suffered convulsions after companies sprayed insecticides near a school. "The companies refused to say what was in the spray," he says. Several schoolchildren also suffered fainting episodes in 2006 and 2007, which were widespread and concerning enough to close the school for several days.[15]

"They were raising seed corn," Bob says. "The wind currents can be thirty miles per hour in one direction and then reverse." He says the companies did not account for drift caused by variable winds and failed to institute appropriate setbacks on their open-air fields. Chemical spray wafted into homes, shops, and classrooms.

"There were parents who picked up their kids from school to be treated, and then at night they'd have to take their parents to the hospital for dialysis," Bob says. Kauai has seen high rates of kidney disease, asthma, allergies and cancer as well as birth defects.[16] "They [the counties] took the vote [on GMOs] and then the judges got bought off."

Not only were fields soaked with chemicals year-round, they were also doused with them in untested combinations. Dr. Lorrin Pang is an endocrinologist who has been a consultant with the World Health Organization evaluating drugs, vaccines, and medical diagnostics. He later returned to Hawai'i, where he is from, and offered his expertise to people concerned about agrochemicals. He pointed out that companies had not tested the effects of multiple exposures:

> When you have two drugs that have been studied and are known and you combine them, all heck could break loose . . . Both drugs could be totally safe by themselves, but when you combine them either one could become more toxic or something new could occur altogether. Well, we're supposed to apply this principle to pesticides. In normal agriculture you might expect 7 or so chemical pesticides being used. In these GMO fields they are using as many as 80 different chemicals and no one knows

what happens when they mix in the fields or drift into the environment and combine.[17]

The threat of agribusiness foisting chemical toxins on rural communities is not going to disappear anytime soon. Dramatic though the consequences of spraying have been in Hawaiʻi, there has been little attention in the rest of the US. Maybe it's because Hawaiʻi is so far away, or maybe it's because the notion of poisons in paradise creates too much cognitive dissonance. But the onslaught of chemicals, the harm it has caused local people, and the indifference of corporate entities and state officials offers a somber cautionary tale.

For many in Hawaiʻi, it is yet another example in a long litany of exploitation. On the first morning I spend in Maui, I take a stroll around the property with Adam. Though the sun has been up for a while, the grass is still moist with dew. Every step I take scatters a wisp of tiny insects. Without thinking, I bend down and collect pieces of black plastic from the pineapple farm days, more than twenty years back. Adam chuckles at my diligence. "You could do that all day for months and still not be rid of it," he says. The plastic was used for mulch to manage weeds and pests, retain moisture, and enhance the effectiveness of fungicide treatments.

Adam introduces me to "Grandma," a sixty-year-old koa, or native acacia tree. She is imposing, ailing in some limbs but otherwise sturdy. He indicates the sloping ravine in front of us and says, "This gulch used to be all koa trees. This spot where we are was among the first eighty land parcels sold in Maui, if not all of Hawaiʻi. So this could be seen as ground zero for colonization." In the 1800s the trees were cleared for sugarcane by the East Maui Sugar Plantation, which predated Alexander & Baldwin by a decade, and later for ranching. (Makawao was known for its *paniolo*, or Hawaiian cowboy, culture.) The ravine now has some trees, but many aren't doing so well, in part because of insect problems. Adam says that Grandma's seeds were used to plant more koa, and so there is a cluster of young koa trees. The name of this town, Makawao, means "where the forest begins."

One positive outcome of the struggle to rein in chemicals was realizing the power of mobilized citizens. Maui became a center of activism, and the base of the SHAKA Movement—Sustainable Hawaiian Agriculture for the Keiki [children] and the Āina—which organized marches and continues to educate people about the risks of farm inputs. The island's engaged citizens are shining that long arc of sunlight on information that many people in power have hoped would stay invisible. For that, they remain a thorn in the side of Big Agriculture and the officeholders seeking to placate the industry.

Mark Sheehan of Maui Tomorrow tracks Maui's environmental activism to thirty years ago when a group was able to preserve Makena Beach, aka Big Beach, for public use. The beach area near Wailea on the south end of Maui had been bought up by speculators, explains Mark. "They thought it would be the next Marriott." After years of watching miles of lovely beachfront turn to hotels and golf courses, many locals decided they'd seen enough of Maui's glistening southern end go to private hands. Saving the popular beach was a ten-year campaign, Mark says. "It was the beginning of the pushback to not give the developers everything."

Mark and others continued to work on multiple fronts, eventually forming Maui Tomorrow in order to be proactive rather than "fight[ing] one rear-guard action after another." The group successfully preserved other public sites. They successfully barred pineapple companies from using pesticides disallowed elsewhere in the United States. And they successfully sued Dow Chemical, which was forced to contribute $4 million to clean waterways damaged by their practices. Mark says the GMO movement kicked the intensity up several notches. The spark, he says, was when Bruce Douglas, the longtime organizer of Maui's Earth Day, woke from a nap with the question on his lips: *Why are we letting Monsanto do experiments on us?* As Mark says, "We are still fighting the plantations in our minds." Douglas's epiphany called attention to the ongoing need to rise up.

I meet Mark, who into his eighties is still slim and active with close-trimmed white hair, at a café in the upcountry town of Haiku. He says he's been alert to threats to Maui's land and spirit since he moved here from his home state of California in 1973. "I camped out at Makena Beach," he says. "In those days it was sort of a psychedelic nudist colony. I thought, *It is so beautiful here that it's going to be Californicated.* And decided that if I live here, I will do everything I can do to protect it."

He says the last three decades have seen a revival in Hawaiian language and culture as the plantation system loses its stranglehold. "Maui is the tip of the spear of the uprising," he says. During the GMO effort, SHAKA and Maui Tomorrow pursued a ballot initiative process that required signatures from 20 percent of all voters. Three hundred people gathered signatures, and a large team organized rallies and community events. Kauia, the epicenter of GMO malfeasance, had similar gatherings.

Mark says the effort "rocked the state." Given the deep pockets, political influence, and aggressive messaging of the agrochemical companies, the GMO moratorium was an "epic win," he says. After the courts knocked it down, he recalls thinking, "*We lost it.* Then we realized that with 23,400 people voting for our initiative, we had a political force—a party of sorts. We decided we had to change the corporate control of elections and run our own slate. No group had ever endorsed a slate of candidates. Four of our candidates were elected. Last year we reprised our efforts and five of our candidates were elected to the nine-person council. We have people coming over from Oahu because there it's a swamp with dollar signs," he says. "Change is going to come from Maui."

The likes of Alexander & Baldwin do not go away quietly. During my visit one point of contention is the fate of HB1326, dubbed the Corporate Water Theft Bill, which would extend the East Maui water rights that A&B had long enjoyed to Mahi Pono. At stake is $62 million that, should such rights not be secured, A&B would be required to reimburse its purchasers. Despite widespread public opposition, business interests, and the legislators many presume to be in their pocket, wouldn't let it go. As Tara Grace of Maui Tomorrow puts it, "A&B is a hungry ghost for water." The 2019 legislative session ended with the bill still in limbo. A&B and its other proponents could not push it through. Mark attributes this primarily to Maui's determination "to throw off 150 years of corporate control by A&B."

The Makawao house is so beautiful that while there I half think I'm dreaming. It is sleek and airy, with a broad deck where you can spend all day watching shifting clouds and the birds: red, white, and black Brazilian cardinals (related to tanagers) and bright red common cardinals. I drift back and forth between the deck and the kitchen and our bedroom, barefoot on the cool floor, admiring the views from each room. The one hitch is that Marco, the resident big cat, feels compelled to show us who's boss. Several nights he defies barred windows and gravity to break into our room, announcing his trespass with a shriek that jars me awake, putting my nervous system on full alert. Finally one night, heart pounding at 3 AM, I wander outside and crawl into one of the two hammocks. I lie there long after the hammock stops swaying, surrounded by soft night sounds and a sky thick with stars.

The house is centrally located and becomes a gathering point for roving agronomists and regenerative agriculture aficionados, and a spot for me to do interviews. And so on Sunday afternoon Leiʻohu Ryder and her partner Maydeen ʻIao show up to talk. Charlotte, who arranged the interview, comes too.

The conversation starts out awkwardly. I tell Leiʻohu that her pule spoke to me, and that I am hoping to learn more about the Hawaiian culture and language in the context of the sugar era's waning. Leiʻohu is quiet for a moment and then says, "I am just one person. I can't speak for native Hawaiians."

Time freezes for an instant and I envision our conversation—our reporter-source tableau—from Leiʻohu's perspective. I imagine that this is all too familiar for her: a white person from off-island wanting to touch the real Hawaiʻi, hoping to sprinkle the fairy dust of authenticity in my book. I can see how this must be awfully wearying. I feel a twinge of embarrassment: I get it. But where to go from here?

In recent years I have tried to begin to reckon with the enormity of the damage—human, ecological, spiritual—that colonialism has wrought across the world. What I learned in Norway helped me see that the exploitation of land by external powers seeps into every area of life, from access to local water now branded and put into bottles to a person's identity and dignity. Imperial societies have imposed the worldview that nature has value only to the extent that it can be turned into products: that forests, arable soils, and wildlife represent "resources." To think otherwise, to see agency and consciousness in nature, is to be primitive.

This is what I grew up with: the idea that European exploration and America's westward settlement represented progress, and that my cohort would become part of this story and therefore fulfill our purpose. I'm ashamed that I spent my childhood in the upstate New York town of Niskayuna ("extensive corn flats" in the Mohawk language), attended Iroquois Middle School, played with friends on Algonquin Road, and sold polyester shirts at Mohawk Mall and never gave a thought to these names and what they told us about the true nature and history of the place. I can't undo my appalling lack of inquisitiveness. I can only do my best to try to learn now what I'd failed to learn when young.

In a certain way Lei'ohu is holding up a mirror to me. I don't want to be one in a long line of those who want a piece of her, to bask in her authenticity. At the same time, unless I engage with native Hawaiians, how can I understand Hawai'i beyond the glossy version a visitor might see? Lei'ohu is a spiritual leader, healer, and educator of native Hawaiian descent. I had been moved by her masterful use of language and her profound love of place and see her as someone I can learn from. And yet there remains the question: *Where does genuine interest end and cultural appropriation begin?*

I still don't know how to answer that. So I share with Lei'ohu and Maydeen the gist of my book and Maydeen starts to nod, to my relief. Maydeen, who wears a green T-shirt that says ALOHA IN ACTION, has a light, bubbly aspect to her. Lei'ohu, in a loose cottony floral shirt, conveys an impression of gravitas. With Maydeen's quiet encouragement, Lei'ohu relaxes and our exchange begins.

"I think we just forgot our relationship with nature," Maydeen says. "Our earliest ancestors knew that without nature there is no life.

It's like your family. *Our* family. When we realize this and embody this, it's a natural flow. For example, our fishermen. They only take what they need." Imagine: An ethic that feeds people and maintains coastal ecologies is pushed aside as it neither creates personal wealth nor adds measurably to the GDP. Instead, right out of the colonial playbook, we get excess production, which means short-term private gain and long-term public and environmental repercussions. I recall the predator culture strategies to dominate the prey culture as Ánde Somby described: Norway's rhetoric of a national "happy family" and its use of legalese, ensuring that indigenous people stay invisible. The wolf has been busy in Hawai'i, too.

Maydeen and Lei'ohu tell me they grew up in Oahu, the most populous Hawaiian island. When they moved to Maui in the 1980s the population was thirty-five thousand. It has since increased more than fourfold. Their families were farmers and fishers, and as children they were surrounded by nature. "They could never get us in the house," says Maydeen. "Lunch was in the trees."

The pressures from outside were challenging, Lei'ohu says, even within their lifetimes. Increasingly, native land was stolen for development and many native Hawaiians were thrust into a situation that left them no choice but to assimilate and take part in the consumer (mostly tourist) economy. For several years Lei'ohu was a teacher. Maydeen worked in a hotel. "We witnessed the transformation. When we grew up it was fish and poi, organic food."[18] Then branded processed foods, introduced and imported from elsewhere, became the staples because they were available and cheap.

Lei'ohu sees the substitution of commercial foods for traditional foods as emblematic of the near eradication of the Hawaiian culture. "This is what happens when the Western world comes in with their views and desecrates what is sacred, including the very sacredness of our land," she says. An example is the word *aloha*, which, she says, Western interests usurped to benefit tourism, depriving the word of its meaning. According to Hawaiian statute, *aloha* "is more than a word of greeting or farewell or a salutation . . . *Aloha* is the essence of relationships in which each person is important to every other person for collective existence. *Aloha* means to hear what is not said, to see what cannot be seen and to know the unknowable."[19]

Indoctrination from the West—from missionaries, colonization, plantation economies—creates insidious wounds. "This happens when a culture is made to feel 'less than,' that they have no value," Lei'ohu says. "Once money is introduced people have to survive, often at the cost of their spirit. I have seen the impact in my own family." However, this isn't where she chooses to focus her energy: "Some of these wounds create mistrust. We've got to get beyond it to use these experiences to lift our spirit, to reconnect to what is true. What is true is that the land is our mother."

In contemporary society we are all trained to "live according to lack: the great fear." By contrast she and Maydeen choose to "take our cues from [a sense of] abundance." She suggests that it is the perception of scarcity and fear, perpetuated by a transactional relationship with the earth, that erodes trust in one another, in nature, and in the possibility of having plenty without retribution: "We're in this lack. It's so hard. With the historical record [native Hawaiians] can say: 'You did this to me.'"

However, she says, holding on to a sense of injustice keeps people from realizing the potential of their land and community. This does not mean she believes local people should simply accept circumstances that disempower them. In the power dynamic of plantation culture, passivity is the flip side of rage.

"Maybe we've got to take our cues from the moon, the wind, the water," she says. She refers to the symbol of the seed, or *ano*, noting that the word means "values," "spirit," and "foundation." Honoring ano informs her relationship to the world. While it is easy to see and dwell on depletion, "I am seeing generosity. We can evolve into the heart way of letting go, to be the open heart of aloha. This takes an authentic embrace of all that is life."

Like John Liu's critique of contemporary economics and Jeff Goebel's consensus work, Lei'ohu makes the case that whether we look at the world and see lack or plenty, difficulty or ease, dissension or solidarity, depends on our belief systems—and that we have a choice.

In turn what we choose, whether consciously or unconsciously, has consequences. If we focus our attention on depletion, this is what will manifest; if we believe a desired outcome is impossible, it will be. Recognizing possibility, exemplified in the unfolding seed, is as much about intention and spirituality as it is rational assessment. I feel humbled by the clarity with which Leiʻohu perceives abundance around her. It is this clarity, this unclouded resolve, that intrigued me. I appreciate that this takes an act of will.

At this time of change in Maui and the world at large, a goal of Leiʻohu's is "being the calm in the middle of the rocky moment. This is the greatest seed that wants to be grown. It is the rejuvenation of the relationship with āina, with all that wants to flourish." In aspiring to "be the calm," Leiʻohu says she has been toying with the word *mana*, which indicates spiritual or healing power that manifests in balance. "Someone who has mana walks into a room and you feel it," she says. "I think: *How do I remain centered in the mana?*"

The Hawaiian language is rich and evocative. Each word seems to radiate outward from its source. I listen carefully as Leiʻohu parses the word *mana*. The *m*, she says, is for *malu*, which means "calm and peace." The *a* stands for *aloha*: love and "the dignity and respect for all life and recognizing the spirit of all life." The *n* represents *nui*, greatness and abundance. The second *a* invokes *akua*, a divine god or goddess.

Calm, love, abundance, the divine. These are meaningful touchstones for anyone. Leiʻohu notes that every culture has language for love. And that Hawaiians are not alone in having suffered from colonization of one form or another. "You can always allow your wound to grow," Leiʻohu says. "It is up to us to choose. Not the seed of hate, historical DNA, trauma, and separation, but the seed of love, of aloha, of connectedness, creating what hasn't been dreamed of yet." She says native Hawaiians have the skills and knowledge to restore the land and grow food ecologically. Local practitioners should not be ignored, nor taken for granted. "These cultures that are blessed to be able to remember themselves are inviting us in," she says. "We have continued to survive. We have schools. Our culture is alive and thriving. The people of aloha are still here." The question, she says, is "How do we get along? How do we create this rainbow? Let's do it now, face-to-face."

———————

Charlotte comes by early one morning and we head out on a road trip to Hana, on the eastern side of the island. The eighty-kilometer (fifty-mile) drive through coastal forests is stunning—and famously tricky, with rough surfaces, hairpin turns, and skinny one-way bridges. We've got snacks and bathing suits and while we settle into the car we watch the birds around us, none of which is native to the island.

"It's always amazing to think of how young this land is, ecologically," says Char. Most birds here are imports, having arrived in Maui in various ways. Char says in 1915 there was a fire in Honolulu, during which Chinese bird keepers set their birds free rather than risk letting them perish. Her favorite Hawai'i bird, the Chinese melodious laughing thrush, could have been among those that were released and made this place home. "It changes its song constantly. It's a brown bird with white stripes, about the size of a robin. Nothing to write home about, but such a singer!"

In Hawai'i, the question of native versus non-native species is complex. Some imported species have caused tremendous ecological harm. For example, sugar companies brought mongoose from India to their fields to control rats. Unfortunately, no one considered what else besides rodents the stealthy weasel-like animals preyed on, and they've taken a toll on the local birds, reptiles, and sea turtles.[20]

Some non-native species fill an ecological niche. For instance, with many native birds under stress from warming temperatures and avian malaria, introduced species are taking on the important role of dispersing the seeds of indigenous plants. Still, the state is extremely vigilant about invasive species, and has instituted early detection and reporting programs. I don't know how the Chinese melodious laughing thrush influences the landscape, but like the non-native Hawaiian people who live on Maui, they are here: part of the visual, sonic, and cultural rainbow.

We drive north through Haiku until we connect with the Hana Highway, moving down and up through heavily forested ravines. "This gulch runs through the old Baldwin property," she says. "The road gets a lot of flash flooding." She says that we will be doing a circle, returning via the south side of the island through Kaupo. This will take us through dry, sparsely populated areas quite different from the lush tropical scenery we've become accustomed to here. "It took me eight years to realize that the dry side is desertified," she says.

That part of the island, up through Haleakalā's leeward side, was once green and thriving. In the late eighteenth century the sandalwood forests caught the attention of traders, as the fragrant tree was a prized commodity, especially in China. People tend to regard Maui's dry side as a frontier, its rugged terrain part of its charm—and assume that it's always been this way. Even Char, with her interest in landscape change, accepted this account until it dawned on her that deforestation left the land arid, as has happened in places across the world. That this took place somewhere as verdant as Maui is hard to grasp. The name *Kaupo* means "rain that makes one hide behind rocks."

Driving south from Hana, our next stop is Whispering Winds Bamboo Cooperative. As mentioned, Char is an advocate of bamboo, a fast-growing perennial grass with many ecological benefits. Bamboo is highly efficient at sequestering carbon; it can stabilize watersheds and decontaminate soil and water. Its shoots are a nutritious staple in Asia. And it is a recyclable, decay-resistant, non-off-gassing alternative to timber and other building materials. In rural Vietnam, Charlotte says, bamboo was employed to rehabilitate landscapes in bombed areas along the Ho Chi Minh Trail. After the war, she says, farmers were given one kilo of rice a day for planting bamboo.

In a 2014 video called *The Power of Bamboo*, permaculturist Geoff Lawton shares how he employs it at Zaytuna Farm in Australia. He says his "wall of bamboo" filters floodwater and captures organic matter and nutrients for use in the landscape. Bamboo creates a living hedge and windbreak, and the leafy parts are fodder for grazing animals. Stalks come in handy as planting stakes, and certain bamboo varieties can be split and woven for use in cob construction. He points, too, to bamboo's strength: "If you're caught in a hurricane, if you're caught in a tornado . . . run for the bamboo."[21]

Char's friend Rich von Wellsheim, who founded Whispering Winds and remains a worker-owner, isn't around, so we take a stroll about the farm and its bamboo forest. What variety! There are shiny, elegant black bamboos, some with deep green canes, or culms, and others with buttery yellow stalks. There is something wondrous about these plants, which can shoot up a meter a day during the yearly "shooting season." Many are surprisingly tall; it feels as if we're miniature people wandering among giant grasses. The canes ripple in the breeze, making a soft, brushy, whispering sound.

Native to Asia and parts of both Africa and North America, bamboo grows well in Hawai'i. Perhaps too well. On our way to the Hana Highway, Char had pointed out a "running bamboo" that had overtaken a stretch of land by the road. "Clumping bamboos" send shoots out from a central base. The escaped bamboo is a runner. Because the runners can so quickly take over, Whispering Winds only grows tropical clumping bamboo.

As we circle back we find that Rich has returned to his modest bamboo-constructed house. Rich has worked with bamboo for many decades. He's an amiable guy, happy to share his bountiful knowledge about his favored plant. "There are fifteen hundred species of bamboo and fifteen hundred uses," he says. "They can be two inches tall or a hundred feet tall. They have tremendous holding ability, useful for roads and for soil and bank stabilization. Bamboo does better as an alleyway or windbreak—not as a plantation." When Char mentions we saw that marauding runner on our drive, Rich shakes his head. "Running bamboo has helped to devastate mature forests here. The rumor is that in plantation days, a Japanese worker brought a bamboo and planted it. It ran. It is to be reckoned with, the bamboo."

Tropical environments like Maui's have high regenerative potential because they are so extraordinarily fecund. Unless you put on the biological brakes with chemicals or chain saws, plants will grow. Bamboos are the speed champions, for sure, but many tropical plants are quick to put on bulk and height. The three trajectories that John Liu highlighted as the pillars of land restoration—increasing biodiversity, biomass, and soil organic matter—occur quickly in warm, moist environments.

Roland Bunch, author of *Two Ears of Corn*, has had great success helping farmers in the humid tropics develop productive systems and restore land with simple, low-tech strategies that build on this advantage.[22] Over five decades Roland has collaborated with rural farmers in places like Belize, Honduras, Guatemala, Brazil, Madagascar, and Cameroon. He advocates using green manure cover crops—often native plants like cowpeas—to fertilize soil and outcompete weeds.

In 2011 Roland noticed Dogon villagers in Mali pruning trees into the shape of a funnel. It was peculiar, but while most people in the region had paltry harvests due to drought, these villagers had a bumper crop of millet and several other grains. With some sleuthing, Roland attributed their success to "intermittent shade." Plants need sun, but in very hot areas too much sun is a liability. Under excessive tropical heat and sun, most crops will stop growing in midday hours. The trees were shaped in a way that managed the shade for the village's food crops. As a local farmer told Roland, "If most of a tree's leaves are high off the ground, their shadow gradually moves across the field as the sun travels across the sky. That way, all the crops, even those right next to the tree trunk, get some good sunlight at least part of the day."

With little fanfare, tens of millions of farmers are growing food using some form of regenerative agriculture, Roland says. He adds that, particularly in tropical areas, farmers are sequestering carbon at a rate far higher than many scientific carbon drawdown estimates.[23] In a series of case studies Roland writes: "Even fairly simple gm/cc [green manure cover crop] systems can sequester more than 5 t/ha/year [five tons per hectare per year] of long-term carbon in decent tropical soil." This means that "if all the world's farmers and ranchers did as well (farmers in the temperate climates won't do as well, but ranchers should be able to do even better), they would sequester about 80% of all the carbon needed to reach the goals of the Paris Accords by the year 2100." The cost is essentially nil, he says; building soil carbon is an added benefit of growing healthy and abundant food.

Many people in Maui, including Adam, are exploring "syntropic agriculture," also known as "successional agroforestry." A short film, *Life in Syntropy*, describes it best.[24] The film introduces Ernst Götsch, a Swiss man who moved to Brazil in 1983 and began farming five hundred hectares (twelve hundred acres) of deforested land with poor, compacted soil. Today the land is a healthy rainforest with a small, productive cacao farm and a more temperate microclimate. The area is cooler and receives more rainfall than it did when Götsch began, and streams now flow throughout the year. Götsch's pivotal innovation is actively pruning trees to build soil organic matter, promote nutrient cycling, and suppress less desirable plants, all of which creates the conditions for thriving, productive plant communities. The "syntropy" contrasts with entropy; rather than losing or dispersing energy, successional agroforestry *concentrates* solar energy. This leads to higher orders of complexity, and increases abundance and resilience to environmental stressors.

Mark Sheehan and fellow Maui activist Jeffrey Bronfman created Haiku Aina Permaculture Initiative (HAPI), a demonstration farm and training site for young farmers. Ben Brown, a twenty-two-year-old intern from Utah, shows us around the upper zone of several project sites. Ben is tall, lean, and tanned from the sun, reminding me a bit of a fast-growing tropical plant himself. I later learn that he and another HAPI intern had a contest to see how long they could go without purchasing any food, relying on what they grew and foraged for. They lasted three months.

We stroll along the meadow path and Ben explains the logic of what's been planted. "Here we're phasing out bananas and papaya," he says. "They grow really fast and easy. Part of the project is that you want stuff that you can eat or sell. This sets up and builds the soil. We'll plant banana every ten feet. Then, say, a cacao every five feet. As those grow and get bigger we will remove the bananas." He is describing how they promote the successional shift from plants that grow easily to more complex plants that mark more mature systems.

Consider the banana for a moment. Banana trees are in the grass family, and their prunings make high-quality mulch. The fruit is actually a berry that grows in rings of bunches above a large, red jewel-like flower. It hangs down from the tree, looking something like an art deco chandelier. Though not native to Hawai'i, bananas have been on the islands for a long time and are prolific here. Ben rattles off several banana varieties grown at HAPI: ice cream (you can freeze it and it tastes like ice cream), apple (the kind we've been picking up at the market), Cuban red, Williams. "Namwa is a favorite," he says. "It's a lot more dense."

Ben points out several other plants right in front of us. Mamaki, which is used for herbal tea. Ti, a sacred plant. Kava, which makes a relaxing tea. Bele, or Tongan spinach, which is rich in calcium and iron. An açai palm, which will offer shade to the nearby cacao. Vanilla bean growing on a stump. A small orange, which, Ben says, "wants to be in shade. Citrus are jungle trees from Asia. They're not supposed to be in full sun." No wonder industrial citrus producers are having a hard time. By planting rows in the open, they are working against nature.

On our last day Char, Tony, and I go to Hōkūnui, a regenerative farming operation with several projects including what they call Poly-Forestry, or Polynesian forestry based on Hawaiian indigenous knowledge. We visit with Director of Forestry Operations Koa Hewahewa, whom locals describe as a genius with plants. He says, "The āina is not just land: It is soil, celestial bodies, microbes, plants, flowers, insects, wind, sand. It is seaweed and the corals. Āina is family. We respect nature as our grandparents. In my house, the sand, wind, and rocks are cared for as our grandparents. When you're from a place for multiple generations, you can recognize patterns, and when we see change in cyclical patterns we are able to forecast events ahead of time. We've been working this land for thousands of years."

This knowledge, embedded in the Hawaiian language, songs, and family relationships, informs Poly-Forestry, he says, noting, "If you're going to be planted in my field, you've got to have a purpose." Often that purpose is feeding the land, mulching organic matter back in. He says he uses "'family-style' planting: elders, cousins, aunties, and uncles. If you are a tree planted alone, that's like a child growing up by himself without family. You don't want to be a vulnerable teenager out there battling the elements on your own."

The Hawaiian planting calendar is based on the lunar cycle. "The Gregorian calendar was created by a monk in a cave," Koa says. "Plants and insects operate by the moon. Why did we detach from it? Each phase has a specific attribute. This connects to the movement of liquid. Not just the tides but also sap in a plant." Another tradition is chanting. Hawai'i is an oral culture, and singing helps people remember traditional means and methods. He says, "I believe that microbes, plants, and soil respond to vibration, tone, and voice."

Koa needs to get back to the field, so we leave his office and have a look at a planting site nearby. "The kalo comes up first, within six to nine months. A lot of food and medicine came out of here," he says. There are trees full of bananas. In Poly-Forestry, as in syntropic agriculture, robust, fast-growing plants are planted alongside more delicate, slower-growing species to offer protection. "This is a windy place," he says. "Pitch a tent in there on a very windy day and you'd be okay."

I am surprised to learn that this tenth-of a-hectare (quarter-acre) plot with dense growth and a wide variety of species was planted just fifteen months ago. The koa and 'ohi'a trees are the linchpin of the system. "Koa trees act as irrigators of the forest and fertilizers of the forest." Regarding "irrigation," the canopies of native forest trees intercept rainfall and fog and distribute moisture throughout the system. Plus, the 'ohi'a tree, a pioneer that is typically the first to grow on lava, has a taproot that enables it to transfer water through blue rock. The plant nursery at Hōkūnui is called Kapū'ao, which means "the womb."[25]

Koa and his team are planting a tenth of a hectare a month. Given the condition of the surrounding land, this involves a lot of work. Bit by bit, they are reviving the landscape with productive native vegetation. Says Koa: "Dig six feet down and still full of plastic. Do we cry? No. We just keep planting."

———

In July and August 2019, fires swept through the Central Valley, scorching thousands of acres of former sugar fields, now Mahi Pono land, including one caused by embers sparked by faulty farm equipment in the fields near Haleakalā Highway.[26] Mahi Pono did not get crops in the ground until the end of August, so the land baked in the heat all summer and brush dried out, providing perfect tinder. Even as the company embarked on its first planting, sixteen hectares (forty acres) of potatoes, eighty hectares (two hundred acres) burned on unplanted fields.[27] Larry Nixon had resigned as Mahi Pono's farm manager in May, just five months after he came on board.

People on the island asked: Is Mahi Pono a bad actor, or merely incompetent? While they talked of growing food for Hawaiians, was buying the land mainly a play for water? Subsequent to the land purchase, Mahi Pono acquired a 50 percent interest in East Maui Irrigation Company, an A&B subsidiary, and the co-owners are seeking long-term leases for water diversions. The fact that the new COO, California water attorney Tim O'Laughlin, specializes in water privatization has raised eyebrows. In a roundup of new Mahi Pono developments, reporter Deborah Rybak articulates what many on Maui have come to suspect: "They are coming for our water. Make no mistake."[28]

Activists like Charlotte, who had been willing to give Mahi Pono the benefit of the doubt, are losing patience. "We've never had fires in the Central Valley," Charlotte says. The sugar companies used to have their own fire equipment, which was likely sold when they liquidated their equipment. One of her daughters was working in Kihei, a beach and tourist area, and had to evacuate. She asks rhetorically: "Who's paying the bill? The county." Char has left Āina First to work on food security strategy with Arnie and Ron Koss, founders of Earth's Best organic baby foods. Āina First is now focused on climate action. There is a continued sense of urgency.

Another development during summer 2019 was the demonstrations against the Thirty Meter Telescope, which was to be sited on Mauna Kea on the Big Island. While proponents say this is a vital location for astronomy research, those opposed to the observatory maintain it is a sacred place for native Hawaiians. Since the 1960s, thirteen telescopes have been built on this sleeping volcano. Many say the fragile ecology of the slope has already suffered through several decades of construction and other impacts like trash and chemical spills. Protestors, sometimes numbering into the thousands, blocked roads to the construction area and managed to press the pause button on the billion-plus-dollar project.

Lei'ohu Ryder, who with Maydeen lent support to the *kai'i*, or protectors, on Mauna Kea, sees the event as a milestone in native Hawaiians' self-determination. Many in Maui interpret this as a sign of Hawaiians' growing resolve about what belongs to the people and a harbinger of what corporate entities who think they can do whatever they want will confront.

Lei'ohu says that for native Hawaiians, who have not always felt empowered, such conviction is coming to the fore. "Aloha is fierce as well," she says. "What is beautiful is that there is a unified voice now. The Westerners wanted us to be disconnected and for a while we were. But not now."

SIX

Busting the Myth

Why Women Belong in the Saddle

I remember in art history classes learning about soil and fertility goddesses. The ancients worshipped these deities because they knew that soil fertility was tied to their harvests and well-being. Somewhere along the line, the sacred, universal connection between women and soil was broken. Now women around the world are reclaiming it.

— Diana Donlon,
founder of Soil Centric

At the New Cowgirl Camp, we don't mess around. After a quick breakfast around a long, hardwood table—handcrafted by our host Beth Robinette's husband, Matt—we're divvying ourselves into cars. We have to rush to get to the cattle area in time to observe the butchering of a cow.

Our group of nine campers and two instructors stand under a sunny, late-summer sky as two men in baseball hats attach a slaughtered cow to a pulley and hoist it in front of the open cabin of a white truck equipped with the tools of the trade. Within moments the carcass is sliced lengthwise in two and the flesh stripped from the ribs—neatly, as if the guy had simply unzipped a zipper. The men hose down the carcass as they toss animal parts into different buckets and on the ground. You can still see the beauty of the animal: the smooth black coat, its lovely mouth.

This is a lot to take in. Clearly, there's no room for squeamishness in this program. "If you don't like it, you probably shouldn't eat it," says my fellow camper Alexandra Machado, who raises goats and produces cheese on two hectares (five acres) in western Washington.

"Ninety percent of our business is custom beef," says Beth, who represents the fourth generation of her family to steward the Lazy R Ranch in Cheney, Washington. She runs the cattle (and now sheep) operation with her father, Maurice, and a part-time apprentice. "You can kill on-farm. It is more humane." The team that harvests the cattle comes to the Lazy R about every other week to process two or four from the herd.

Beth wears a red-and-black-checked flannel shirt and jeans, her blond hair drawn high and back. She laments the lack of markets for perfectly good cowhides. The processors are no longer taking them, she says. "The supply chains haven't been worked out. Like most parts of our food system, it's disconnected." At the Lazy R, she says, "we're using as much of the animal as possible. The hooves are used in gelatins. Organ meat has become more popular. Tongue is popular in the Hispanic community. Heart is not as popular, but some people will take all I've got." Indeed, a short while later a man shows up to collect the tripe, or lining of the stomach, from one of the discard pails. It is the size of a small throw pillow and looks like honeycomb.

Our co-instructor Sandra Matheson says slaughtering has risen in price: "It costs me $2.50 a pound to get the meat USDA-wrapped." A retired veterinarian, Sandra raises cattle for beef and breeding in the far northwest part of the state, sixteen kilometers (ten miles) from the Canadian border. She also has yaks, and was among the first in the United States to successfully breed yaks via artificial insemination (AI). Compared with beef, she says, "Yak has a mild flavor. It's leaner and higher protein." She recently received a USDA grant to develop yak sausage, which she says we will try later in the week. The animals, native to Tibet and the Central Asia highlands, produce both wool and coarse hair "as fine or finer than cashmere." She says

yak babies are cute, but that "a frightened yak is a dangerous animal."
I don't doubt her.

The cost-of-processing conversation segues to genetics, a hot topic
among ranchers. Sandra's cattle are a Simmental/Angus mix. "I select
genetically for marbling on grass," she says. "I look for natural fatness.
We usually butcher at eighteen to twenty-four months."

Beth tells us the Lazy R has lowline Angus genetics: more like the
classic Angus of the 1960s with a smaller body frame so that the animal
puts more energy into muscle, fat, and ultimately flavor. "Now I'm
trying to get the carcass size up a bit," she says. "I'm selecting for docil-
ity. That is why I'm partial to Maka." She gives a firm but affectionate
nudge to a stout bull with a fine, brown coat the color of extra-dark
chocolate. "He is a really chill dude. You should always respect a bull,
even if he seems tame. It's important to follow some principles for how
to interact with the cattle. We don't keep dangerous animals. We pay
attention to how they move down the chute [for artificial insemination
and to check pregnancies]. If I have a heifer who's wild and snorty, she
is not replacement heifer material. Every time you work with them in
the corral, if you see aggressive behavior you mark that down."

Each of us, leaning against the fence or snapping photos with our
phones, takes in every word. All these details matter: the processing
and supply chains; maintaining herd genetics; how to assess snorti-
ness and danger. For this weeklong workshop is more than a lark, a
chance to hang out in a scenic western landscape with like-minded
women. For most of the group, it is an initiation to a career working
with livestock on the range.

In Beth's words, "We are assembling an army of cowgirls to change
the world."

———————

Women have always played an important role in farming and ranching. But in the United States and other Western countries, there is
growing interest in agriculture among women—that is, women are
actively choosing this path as opposed to landing here through family
circumstance. More than half the farms and ranches in the United

States are likely to change hands over the next twenty years, and a new generation of female farmers is poised to step in. According to the USDA, 31 percent of the country's farmers are women, and this cohort manages upward of 120 million hectares (three hundred million acres).[1] In the Midwest, women own or are co-owners of close to half of all farmland.[2]

The number of farms managed by women stands to rise for several reasons. Farmers' children are often not inclined to work the family land, which frees up opportunities for others. Or a daughter rather than a son may choose to take over the farm. Women live longer than men, and some women find themselves running an operation after a husband dies. Also, women are traditionally charged with feeding the family; the sustainable food movement has inspired many people to question where the food they buy and serve is coming from. Producing food is a way to nurture people, animals, and land, and is for many a route to empowerment. New equipment, such as lightweight portable fencing, makes moving animals quicker and easier. Finally, more and more people see farming and ranching as a means of addressing environmental concerns, which attracts women seeking livelihoods that make positive change in the world.

Sarah Wentzel-Fisher is executive director of the Quivira Coalition, an organization devoted to resilient western working landscapes. She says their apprenticeship program has historically drawn more women than men, though it is starting to even out, and adds that the strongest applicants have tended to be women. "We have some great success stories, women who jumped into leadership roles and stepped into big operations," she says.

As for restorative approaches in particular, she says, "I think there is something about regenerative agriculture that requires a kind of holistic thinking that is more socially aligned with the kind of unconscious cultural training that informs women in the US." In our culture, women are geared to be attuned to emotional and other signals. This creates "a skill set you need to have to practice agriculture in a way that is thinking about the whole, the processes and the system," she says. "Increasingly, there are young women who desire to have physical kinds of jobs and develop a set of technical skills and regenerative agriculture provides a space to do that. We're starting to see this and at the same time we're not there yet. There is still a lot of inherent sexism that comes from traditional agriculture in rural communities. Women of all ages bump up against that."

Julie Sullivan co-runs San Juan Ranch, an organic, grass-fed beef operation in south-central Colorado, with her husband and partner George Whitten. Their apprentice program gives her a chance to meet young ranchers, most of whom are women. She told the *Colorado Springs Gazette* that during the interview process, the women seemed eager to learn while "the young men sort of felt like they needed to show us what they already knew." She continued: "I think young men are raised in our culture to feel like it's not okay to say when they don't know what they're doing. They feel like they're supposed to figure it out for themselves. Whereas these young women, for whatever reason, feel really comfortable saying, 'I'm fine driving this truck but I'm having trouble shifting with this one. Can you come show me?'"

She indicated that the culture of their ranch, which integrates agriculture and environmentalism, dovetails with what many would-be women farmers are looking for: "We really do consider our relationship with our land and our animals to be one that is about reciprocity and one that's about learning from the land and the animals as much as we're asking it to do something. That thing about deep relationship really appeals to the young women who want to go into agriculture."[3]

For some women the lure is, simply, soil. Here I am an outlier: I came to an interest in soil from learning about it, and appreciating the extent to which soil is a crucible for so many global problems—and a crucible for solutions. But I can't tell you how many women I meet who say they have always felt an elemental connection to soil, and how it "grounds" them. There is a group of us that regularly meets on Zoom, six women drawn together by our interest in soil and its centrality to our work. We are from Australia (Cindy), Zimbabwe (Precious), California (Diana), Minnesota by way of Montreal (Alex), and Vermont (me and Didi, though in different corners of the state). We share news of our projects, test out crazy ideas that sometimes amazingly come to fruition, offer support in numerous ways, such as encouraging one another to ask for fair compensation (a frequent theme).

Weruschca Kirkegaard, who runs the Dutch-based regenerative consultancy Mūla—Ecosystems by Design, says she finds working with soil "healing and empowering." (*Mūla* in Buddhism means "origin" or "root.") She and I have become across-the-world friends, and I've cheered her on as she's racked up accreditations and honors. In a recent chat she shared with me the path that brought her to permaculture and ecological repair. She had worked in interior design, specializing in antique building materials. Around the time she started a restaurant, a relationship crumbled, she hit a financial wall, and she began to suffer from fibromyalgia. "I fell into a ravine," she said. "I got very sick. I was told I should prepare for my life to be in a wheelchair. I was in bed for about one and a half years. The only thing I could do was read."

Weruschca delved into books on nutrition and health. "The more I applied it to my own life, the better I got," she says. Her interest in food broadened to the growing of food and then to nature and biodiversity. Once recovered enough to leave the house, she returned to school and started volunteering in a vegetable garden. She says: "I felt I had fewer aches when I was in the soil. I went barefoot. It was literally making me better. It was humbling, to have a love affair with the soil. Then I discovered regenerative agriculture and permaculture and the work really opened up to me."

In many ways, Weruschca says, this was a process of coming home, as her father was a farmer as well as a commercial pilot: "He owned a farm with a large nature reserve in Denmark. We spent days chasing and counting insects, amphibians, and mammals like foxes."

She says a parallel journey to her own healing has been a growing awareness of the "brokenness" of our current food and economic system: what it has done to the environment and the role of soil in healing the earth, place by place. "I wanted to be part of this change in whatever capacity I could," she says. "Life has taught me the hard way that no relationship is certain. This work is one relationship I'm confident in growing old with."

I wasn't even sure cowgirl camp would take place. On August 7, 2018, just a few weeks before the start date, I saw a Facebook post that began, ominously, "One of our worst fears became reality last night." The ranch is adjacent to I-90 heading toward Spokane, and the day prior sparks from a vehicle on the highway ignited several fires in the area. Fire crews and BLM worked through the night to get the blaze under control.

Distress notwithstanding, Beth was grateful. She wrote: "All told we lost about 100 acres but all structures, livestock, and people are safe." The photo was a sepia image of what must recently have been green: brown earth; twiggy trees stripped bare three-quarters of the way up; a band of smoky gray. I felt my throat constrict just to look at it. I had met Beth several times at conferences. This was not happening to a stranger.

I registered my concern and sympathy and privately wondered how she could possibly bear the thought of people showing up at the ranch in the aftermath of this fire. But the very next day, I received a straightforward pre-camp note about logistics and asking our T-shirt size. Okay, then, the show must go on. I supposed if you're going to ranch, you have to be prepared for anything. But I continued seeing news reports of ongoing fires in the Pacific Northwest, and one day Beth posted that Spokane had the worst air quality in the country. What would I be in for? It seemed Beth hadn't thought twice about forging ahead. Far be it for me to be intimidated by a bit of smoke.

As our plane approaches Spokane, I see what looks like water droplets on the window. There's smoke but also clouds, many tiers of gray. The woman at the airport information desk apologizes to me about the air quality. I tell her it is raining and she brightens. How gratifying to be the bearer of good news.

Beth picks me up at the airport. My tent mate, Megan Meiklejohn, is already in the car. Megan is the sustainable materials and transparency manager at Eileen Fisher. This means, apart from the fact that she spends her time surrounded by elegant clothing, that she scouts out socially and environmentally responsible (ideally, regenerative) fiber for fabric production. The quest to locate responsibly produced wool and cashmere has taken her to farms in South America, Australia, New Zealand, and Mongolia. And now to Cowgirl Camp, for a deeper sense of what animal managers do and the chance to take the reins.

En route to the Lazy R, Beth points out where the land burned, which is visible from the highway. "One hundred of three hundred total burned," she says, matter-of-factly. "We've lost more trees than I thought, but it could have been much worse. Ponderosa pine is all that grows here." Landscapes dominated by ponderosa pine are prone to frequent fires.

Unlike other campers who live close enough to pile their equipment into cars, Megan and I flew from the East Coast. And so Beth's parents have generously lent us their tent. Beth twists and notches several light-as-air bendable poles and we have a nice shelter with more zippers than I can count and a polyester divider for privacy. Our cluster of tents is right in front of the house; the setup, naturally, is determined by where the sheep will be grazing. We are part of the rotation, and the Dorper sheep get first dibs on the real estate.

While we wait for more campers, we wander over to visit the flock. This is Beth's second season with Dorpers, a South African variety that doesn't need to be shorn (hair sheep) and thrives in dry regions. Plus, they go for types of forage that cattle will turn their noses at. These sheep have white bodies and black heads and necks. We meet Peabody, who was frail at birth and so was bottle-fed in the house. Needless to say, Beth grew attached to him, so likely he'll be sticking around as opposed to sold as breeding stock or for meat. Possibly he'll end up as a "ram-panion": a buddy for a breeding ram, who, kept away from the ewes and lambs, has a dull and lonely life when he's not "working."

Beth is tall and strides easily among the animals, pausing here and there to adjust a fence pole or nudge a frisky lamb. We're at the edge of the Palouse, a sprawling prairie of undulating hills where the dominant crop is wheat. It was formed by loess, or silty soil blown from the south, similar to that on China's Loess Plateau. Maintaining healthy, diverse grassland here is important to Beth. "Most people have sold their land in this region," she says. "There are one hundred houses on this road that weren't here when I was growing up." Spokane is growing, and because this spot is near the city, development has encroached.

She says she chose to put together these camps—the New Cowgirl Camp debuted in 2017, and there's a co-ed New Rancher Camp that Sandra hosted earlier this year—because "I'm getting over the idea of trying to convince people to change how they do things. It's better to give people who want to make a change the tools to get started." She says she would explain holistic planned grazing to area ranchers "and they'd reply: 'That's interesting. But it would never work here.' At the same time, when I explained to friends they'd say, 'Oh, so you're trying to work with nature.' I started asking: *Why are we spending all this energy trying to convert people who don't want to be converted, when people are hungry for these ideas?*" Particularly, she found, many young women.

We drive into downtown Cheney—pronounced *CHEEN-y*, home of Eastern Washington University—for Mexican food, and then Megan and I settle into our five-star tent. It rains softly all night long: a soothing *paddle-paddle-pop-pop* sound as drops of water ping the roof.

After our encounter with the processors, we gather back at the house. We chat in the living room as a robot vacuum skitters around, occasionally bumping into walls. The two dogs of the house, Phoebe and Mayday, cruise around seeking strokes and adoration. Everyone admires Beth's tattoos, particularly the one on her upper arm: a cowgirl riding a cow with ponderosa pines in the background. She establishes the tone for the week when she declares this a "low-rhinestone zone." If you're in it for the bling, you've come to the wrong place.

We do a "grounding," in which each of us shares what brings us to Cowgirl Camp. Ruby, twenty-three, begins. Her eye is swollen, the result of an early-morning run-in with some yellow jackets while she was moving the sheep fence. "I want to learn how to start a farm and run a farm," she says. Despite her youth, she does have experience. She grew up on a ranch run according to Holistic Management in upcountry Maui, probably not far from where Tony and I stayed. After graduating from Dartmouth she spent several months working at Fjällbete, a sheep operation in Sweden that's a cornerstone of the Savory Institute Nordic Hub.

Megan Meiklejohn says she hopes to be able to have "deeper, more intelligent conversations with the farmers we meet. I'm always that young person from New York City who has no farming experience."

Amanda, a photographer based in Oregon who works for the *New York Times*, says she has long been fascinated by women farmers, inspired in part by some land her family owns. "We're at the edge of a precipice with farming," she says, adding that she's interested in "opening new ways of doing agriculture where we are helping each other."[4]

Cielo, a hospice nurse originally from Colombia, tells her story: "I wanted chickens. My economist husband said, 'For how much it costs to produce a dozen eggs at home you can go to the grocery store and buy three dozen.' I built the coop myself. Then I said we need a homestead. He said absolutely not. We ended up buying a homestead. Then we bought sheep. All of a sudden this country-clubber guy with a fancy car has a truck. Everyone's idea [of success] is to move from the country to the city. And I was going back to what I had been trying to escape."

Sandra says as instructors their goal for the week is to impart knowledge and confidence. She says that one of last year's campers was able to save a calf thanks to what she learned here. When the cow exhibited certain signs, she knew to "scrub up her arm and how to assess what was going on. The calf was backward. At 2 AM the vet came and saved both the cow and calf."

Sandra turns to a fraught topic: the farm mascot, an animal, perhaps like Peabody the lamb, that a rancher simply cannot part with. "It's okay to have that kind of emotional attachment," she says. "Just because you're a farmer or rancher doesn't mean you have to have no feelings. You care about your animals. It's hard when I butcher our cows. I cry." If you're in the business of animal stewardship to provide nutritious, health-giving food to people, "you can't have all pets. Having that higher degree of compassion makes us better farmers and ranchers." She says she farms the land where her father raised beef cattle and that it's important to her to sustain that legacy.

She reads from an essay she wrote about the life of a rancher: "It's saving the one who couldn't have made it without you. It's crying with the mother for the one who didn't make it. It's a 24-hour job. If I don't care for the animals, who will?"

"Blood, guts, and tears before lunch," Beth says, giving the vacuum bot a little kick. "There are all these people out in the world who want to do this and if you haven't grown up like I did, how do you start?"

Now we veer into a subject that has lurked unspoken: the unique challenges faced by women ranchers. Beth says that despite their equal partnership in breeding and rearing cattle, "people would talk to my dad and not me."

She recalls a ranching conference a few years back where she felt the many women producers were diminished by the attitudes of some men. She says a male speaker said "how 'we need to acknowledge all the women who make this possible.' I thought: This is bullshit—and I know a lot about bullshit. It took me a long time to have confidence. Can I do this by myself? Did I make a mistake by not finding a guy who does this?" Matt is a teacher and, in Beth's words, being a rancher "isn't his jam." She says: "You don't have to be a big, strong dude to do this. We need to create a space to give women the tools to manage land. Agriculture is going to change in the next ten to twenty years. Food production is now such a white-male-dominated field. But just like in a plant community, we need diversity in management."

Katharine, from Montana, says, "It's super inspiring to have women here who don't have partners involved." She says she's felt that even considering ranching on her own was a "pipe dream."

Most of the women ranchers she's photographed were "unpartnered," Amanda says. "Either they're choosing not to be partnered or the work causes conflict."

Beth says that while the lack of respect for women sometimes gnaws at her, "I can't be bothered by that old paradigm because it's slipping away. I'm running a fairly successful business while [other] ranches are going under."

When it comes to misogyny in agriculture, says Sandra, "sometimes you let it go, sometimes you need to stand up."

Alex says that social media "is my enemy as a female farmer. It leads people to compare themselves to others."

Beth, who is active on Instagram, says she would rather people share the hard stuff as well as the cuddly baby lamb pics: "I always post when an animal dies and I feel like a failure. Once a heifer had a wire wrapped around her leg. I posted it. People appreciated it. I was surprised by the response."

Danica, twenty-one, of Spokane, asks: "How can we make the rest of the world respect us if we can't respect each other?"

———

In the United States and other developed countries, women are entering farming and ranching by choice for many reasons: to provide healthy food for their communities; a desire to heal the land; a love of the work itself. In much of the world, however, growing food for markets and home consumption is the responsibility of women, usually among many other responsibilities. According to the Food and Agriculture Organization of the United Nations (FAO), in developing countries women represent an average of 40 percent of the agricultural workforce, and much of this work is unpaid.[5] In some of the world's poorer regions, women produce 60 to 80 percent of the food crops, primarily on smallholdings of five hectares (twelve acres) or less.[6]

Natalie Topa understands the challenges of women trying to produce food under difficult circumstances. She works for the Danish Refugee Council, helping to build resilience and livelihoods for displaced affected populations in the region she's responsible for: ten countries in East Africa, as well as Yemen. The populations she works with include internally displaced persons (IDPs) as well as the communities where displaced communities are settled, often referred to as host communities.

I catch up with Natalie one morning, which is late afternoon her time, in Nairobi, Kenya. Skype and phone calls, some of which she needs to take, are streaming in. She apologizes and explains that she's coordinating a training program in Somalia and in four hours colleagues are due to land in Mogadishu en route to Dolow. At 4 AM the next day, taxis would show up to take her and other Nairobi-based colleagues to the airport. And visas had yet to come through.

"There are only three flights a week to Dolow," Natalie says. Dolow, sometimes spelled Doolow, is in south-central Somalia near the Ethiopian border. It is an area often buffeted by famine, drought, armed conflict, and outbreaks of infectious diseases like cholera.[7] "If we don't make tomorrow's flight, we won't get there until Sunday and consultants need hostage training and security training to prepare." She chuckles, as one does when things are beyond one's control. The situation seems not to faze her.

"I'm always moving a lot. Usually one out of every two weeks I'm in a different country," she says. "I'll be in Somalia for two weeks, then have one day at home. Then to Ghana for a conference and Burundi for an assessment." In Somalia the focus is on designing for resilience—this is where Natalie's background, which is in urban planning, comes in—and permaculture training, highlighting what's especially relevant to camp settings. "We'll meet on a farm and look at soil building, earthworks for passive water harvesting, integrated design of small and large spaces, integrated waste management, biofertilizers." She rattles all of this off quickly.

In comes a call, which Natalie dispatches in about thirty seconds and then tells me, "The government regulations changed last night. We risk being deported if we show up without a visa expecting to get the usual visa on arrival, and Mogadishu is a crazy place. We'll do security training on Saturday in Mogadishu and Sunday proceed to Dolow."

Through the connection I hear the unmistakable Skype ring, and she takes the call. After a minute or two she turns back to me on Zoom. "Well, looks like I get to sleep in tomorrow," she says. "There will be guards with us. In Dolow, we'll have ten armed guards, like a micro-army. We'll be digging and I'll be covered and it will be hot."

Covered? Of course, Somalia is a Muslim country. I'm trying to picture her there. "Covered, because of the culture?" I ask. "You can say culture or you can say patriarchy," she says.

The question of the trip resolved, Natalie relaxes and tilts her screen to give me a virtual tour of her apartment, which she calls a "fifth-floor farm." "We've been making mushrooms," she says. "I have chickens on my balcony." She points out a banana tree, vermi-composting bins, mealworms and beetles, Malabar spinach. I later find a social media post of her announcing "the Grand Opening of Buggingham Palace": an eco-hotel crafted from a crate, brick, and local waste and recycled material to attract beneficial insects. It offers "Free Wi-Fly." For all the poverty and privation she deals with every day, Natalie manages to have a good time.

The "Topacabana," as she calls her home farm project, aligns with her mission of providing tools for people striving to meet their families' needs while displaced. "I'm trying to show what's possible on a tiny scale," she says. The refugees she works with may be impoverished and have trauma, HIV, or other challenges. They might not be in a position to do a lot of physical labor, she says, "but at least they can grow food." She brings the techniques and ethics of permaculture into her work, including building efficiency, reuse, and recycling into the design: "When any energy comes in the door in whatever form—as chickens, old newspapers, et cetera—I like to give it as many spins on the merry-go-round as we can by cycling the waste in compost and food for soil for my garden. I want waste materials in my apartment to exit through the toilet instead of the garbage can.

"We now have seventy women growing perma-gardens in the IDP camps in Mogadishu, Somalia. They were trained by our team, who learned in our Resilience Design Training. Some have harvested five times already. This means they're saving money on food and also have vegetables to sell. Women are now able to use these savings and income to afford school fees for children. Having a garden is like having a 3-D printer for money."

Women's propensity to grow food is not even in question, she says. "It's a mythology that farmers are men on tractors. More than half the farmers in the world are women, often with babies strapped on their back." She says that women have a greater inclination toward agroecology: "Women need different plants at different periods for different purposes—the three *P*'s. They are saving seeds. They need to buffer the conditions in their lives. Men are more likely to move toward using the family's whole plot for products for regional and global markets, which may involve clearing the land." She suggests men are wired for "the long con. This decreases resilience."

Ecosystem restoration has particular bearing on women, Natalie says, because land degradation heightens exposure to risk, including domestic violence. She offers a hypothetical example in Karamoja, an unforested region of Uganda where she once managed a project. "There are pastoralists and cultivators. The land dries out every year, and so pastoralists have to move their livestock. Now when the rains come the roads flood." Let's say a truck is going to Moroto town carrying sugar and oil and pulses, she says. With roads flooded the driver has to make a large detour, so by the time he reaches the rural outpost the prices have increased. "A woman goes to the market. Either she buys smaller amounts or she buys fewer products. Either way, there is less food on the table. The woman's role is to put food on the table. If her crops have failed or she can't afford to buy enough, she'll be beaten by her husband."

Land degradation strains families in many ways, says Natalie, including provoking "crises of masculinity." If cattle die and the family can no longer pay school fees, a man's frustration and loss of pride "can manifest as domestic violence. When there's a drought or other problem, the first one to get pulled out of school is the girl. Then there's a series of trapdoors she may fall through." In pastoral societies, she says, "We can draw a calendar of risk windows. In the dry season, a girl is likely to get pulled out of school. She may be asked to collect fuel in the bush, where she is exposed to risk of sexual violence. In times of plenty when cows and camels are fat and pockets padded, men are looking for a wife. Girls are always up for grabs for different reasons. If there's a drought and men are out cattle raiding, women are left to fend for themselves and are vulnerable to gang rape."

Natalie's perspective is informed by growing up in a low-income home with a single mother who was an immigrant to the United States, she says; she understands the struggles and feeling of invisibility experienced by many of the people she now works with. "I feel the whole world should be designed for people like my mother: that there should be accessible transit, affordable housing, dignified employment, safety and security, not a food desert, places to recreate and play with her children." The ethical design orientation of permaculture appeals to her, and seems increasingly relevant to her work. "Picture it: A woman can meet almost all of her needs, and what can't be met she gets through sharing with neighbors. It allows us to hit the reset button on how to do human."

For the training in Somalia, Natalie has asked for a gender balance among participants, though making that request is no guarantee. "Sometimes there may be thirty men and two women," she says. "I will say it's not okay that men are getting these powerful positions. Men and women both have to own this transformation. Women do a lot of the heavy lifting: water carrying, stick bundling, growing the food. Let us have the same balance."

———————

On day two we learn how to castrate a male calf.

Beth, wearing a straw cowboy hat and her hair in a braid, stands in front of her parents' farmhouse, pulling metal implements from a plastic crate one by one. "This is a big elastrator, for big old nuts," she says. "You have the animal in the herd gate. You manipulate the testicles—give the elastrator a nice swift tug when you feel you have both testicles underneath. You want to castrate as low as possible. Basically, what this does is cut off the circulation to the testicles. Eventually that happens and they fall off." We're sitting comfortably on the grass as she gives the demo. "It's basically like a slingshot. You have to pull the rubber band off and draw back until it's nice and tight. And clamp the ring."

"Make sure you don't get your finger caught in it," Sandra adds.

Sandra, the artificial insemination expert, leads us back to where the cattle are and shows us how to palpate a cow to see if she is pregnant. She puts on rubber gloves. "You can use lubricant or spit," she says. "Take your jewelry off—so you don't hurt yourself or the animal. Work your way in through the anal sphincter. Make a fist and work your arm in. You can feel the uterus. And the cervix: There are rings on the cervix like a pipe. At thirty-five days is the earliest that you can feel a fetus." Her arm disappears into the cavern of the cow's body. She says, "Cows have strong sphincter muscles. You can get bruises on your arm."

When she returns home to her farm, Sandra will be turning her attention to breeding the yaks, initially watching for indications that an animal is in heat. She says that signs include "bumping heads with each other. Or I might see animals try to jump the cow in heat, or the one in heat will jump. With the beef cows, if I see her in standing heat in the morning, I will breed her in the afternoon. Some people rush it." The term *standing heat* means "ready to receive the bull." Although in this case, rather than an actual bull it will be the pipette in Sandra's hand: carefully curated bull semen that sells for twenty-five or thirty dollars per unit, or "straw."

Back at Beth's place we sit on the lawn and try out our animal care skills. We practice giving vaccines, using a clementine as a proxy for a cow. Sandra says, "When I was younger, I once lost twenty-five calves out of ninety. That cured me of not giving vaccines."

We also do an exercise with an "emasculator," which is used to castrate smaller animals. It looks like a big pair of pliers. Sandra hands us each a pair of balloons filled with sand to approximate goat or sheep testicles. The tool's pressure crushes blood vessels so it doesn't bleed; then the testicles fall off. I put a rubber band on the metal piece and squeeze. The sand-balloons fall neatly into a small mesh bag, and I now have a souvenir of Cowgirl Camp. I note a label on the package that reads: "Not for human use."

We have sessions on business planning, holistic decision making, when to call the vet, and selecting a herding dog. Charming and fetching as they are, Mayday (an Australian cattle dog) and Phoebe (a McNab shepherd) are useless herding-wise, Beth says. Phoebe was a rescue. "My guess is someone got her for a working dog and realized she's kind of a wuss." And so we watch videos of clever, fearless border collies maneuvering cattle a hundred times their size into a corral, and imposing livestock guard dogs that protect against predators. Of the guard dogs, the Turkish kangals are huge and on the aggressive side; Maremmas and Great Pyrenees are handsome with white coats.

We also learn low-stress livestock handling. Beth points out that "animals are incredibly sensitive to anxiety and stress and will sense when you're intimidated or frightened." The central concept in minimizing animal stress is the "flight zone," which you can think of as the animal's personal space. "It's how close you can get without them running away," says Beth. It's most effective to position your body right at the edge of the flight zone.

"I do a lot of communicating with animals through body position and body language," she says. When her father, Maurice, moves animals, "there's a lot of whooping and hollering. I do it a lot like natural horsemanship, where it's about how we apply pressure and relieve it. It's a paradigm shift. The old way was, *I'm going to force that animal to do what I want it to do.* That's changed to, *I will let that animal do what I want it to do.* You want to set up every situation so our ideas become the animal's idea."

She lets us in on some insights about what makes livestock tick:

- Animals want to continue in the direction they are already headed.
- They want to be in a herd or flock.
- They want to move away from any external pressure they are feeling.
- Good movement attracts good movement. ("It's the tai chi of the cow.")
- They want to follow other animals (so you just need the first one to walk through the gate).

Sandra notes that cattle and sheep have poor depth perception. "If they see a shadow, they may think it's a crevice," she says. "I've seen animals leap over shadows."

We learn some simple livestock handling techniques. By walking in straight lines, you can stay at the edge of the flight zone. Walking back and forth in a zigzag configuration mimics the instinctual interaction between prey and predator. Moving forward parallel to the herd will slow down movement, whereas going in reverse will speed forward movement. If you stop moving, the animals will look at you—and may get distracted.

Time now to go out and move sheep. "My preferred pace is a mosey," Beth says.

———

On Wednesday we go to the site of the fire. As we walk up from where we park I feel self-conscious, as if I were intruding on the scene of an accident. I can see the remaining singed trees, and, on the forest floor, the ashen duff: a mat of charred material jumbled up with straw and brown dirt. Yet past the rubble there are pines standing green and tall. It's a fine late-summer day with lofty clouds. The sky is blue and the air clear.

"It smells sweet—there's a weird sweetness to it," says Amanda. She sighs. "I've smelled so many fires."

We stroll quietly to a clearing, which I realize might not always have been a clearing. Beth and Sandra look down at the ground, poking around the burnt litter with their boots. We had just talked about soil monitoring and the mineral cycle. Fire is one way to release nutrients. Creating the conditions for nutrients to cycle through ruminants, soil microorganisms, and plants—so that organic matter is decomposed biologically instead of chemically—is another. This was the original plan, which the fire has disrupted, for now.

"There were several vegetative fires at once, and the wind kept changing direction," Beth recalls. "We did backfires, which got a little out of control. We had six-foot flames. It was hot. The smoke was pretty brutal. The firefighters were great—they went to statewide mobilization within four hours. We had firefighters here for a week. I was pretty excited about it as a learning opportunity—and was feeling pretty good until I realized we lost more trees than I thought. The trees here are pretty resilient, but what did burn got really crispy."

I am struck by the nonchalance with which Beth seems to have reconciled this recent calamity, even welcoming what it may teach her about better land management. No one can stave off all disaster, large or small. But you can make the best use of circumstance, chosen or otherwise.

During the fire, "it was awe-inspiring to see the raw power," she says. "Also, I'm fascinated by how the land responds, and how that is determined by how the land was managed." Before western settlement this land was managed by indigenous people who often used fire as a tool. "My goal is to re-create the old-growth forest savanna. I'm selecting for the big, strong trees. Maybe when I get to the end of my fifty seasons here, it will be a ponderosa parkland. I am managing a treescape and landscape that will last well beyond my lifetime. This gives you clarity of purpose."

The area had been logged three years prior to reduce the fire risk. "We had already removed some of the smaller, weaker trees," Beth says. "We hadn't grazed here this year. That would have removed some fire fuel. But then the cows weren't there when the fire happened and I'm pretty happy about that. My hunch is that the grass will come back pretty well next year. I'm not sure whether it will be grazed next year. At this point we're still assessing damage. We'll see what the land tells us. We're already getting a little bit of regrowth where there had been moisture. There's still ground litter. I hope we get snow and not torrential rains."

———

With destructive fire seasons increasingly hammering the western United States, there is growing interest in holistic planned grazing as a tool for managing fire fuel. This is compellingly shown in an eight-minute film, co-produced by Fibershed, a California-based organization that promotes ecologically sound, regional fiber production, and Soil Centric, which promotes opportunities in regenerative agriculture. In it, we meet Jean Gowan Near of Utopia Ranch in Mendocino County who, at 104, still rears Merino sheep prized for their fleece. She says she originally acquired sheep for fire control, fifty-three years ago when she retired from a teaching career.

After the 2017 Redwood Valley Fire, Jean Near's son and daughter-in-law, John and Carrie Ham, who also raise sheep, got a lesson in livestock as fire control. In the video, Carrie points out the difference between their land and that of their neighbors, who do not have animals. Their land looks scarcely touched while the neighbors' land has broad swatches of coal-gray and groups of charred trees. She says, "If you've got a tree that's hanging down, and tall grass underneath it, the grass acts as a ladder and [the fire] just goes up the tree and you lose the tree too."[8]

Ever since I started writing about Holistic Management, I've been impressed by the women in the field—their number, their apparent ease in what would be considered a man's domain, their personal style. The CEO and co-founder of the Savory Institute, Daniela Howell and Jody Butterfield, respectively, are women, as is the executive director of Holistic Management International, Ann Adams. I've met women from across the country and around the globe with a deep commitment to the people and wildlife of their region and a passion for sharing knowledge. They seem so comfortable in their own skin, perhaps in part a consequence of working with animals. When I'm around them I get a certain spark, that feeling of *This is what I'd like to be when I grow up*—though I'm older than most of them.

One group of practitioners is young female graziers, roving shepherds with movable flocks who help private and public property owners improve their land. In the West an important component is minimizing its flammability by reducing vegetation that could pose a fire hazard, something herbivores are uniquely equipped to do with neither chemicals nor noise, apart from the soft sounds of chewing and the occasional bleat.

Los Angeles–based Brittany Cole Bush describes herself as a "modern day urban shepherdess, shepherding animals, people and

projects."[9] In addition to contract grazing (the livestock management version of freelancing), she makes custom leather products by
hand, as a partner in Shepherdess Holistic Hides. She is also helping
to develop the Grazing School of the West, a vocational training
program for regenerative rangeland management.

For several years Bush managed prescribed grazing programs in
Bay Area regional parks to curb fire-hazardous vegetation where wild
lands met the urban edge, and to contain invasives like poison oak
and blackberry. "These [grazed] areas were not affected by the recent
fires," she said in a 2018 webinar on post-fire management.[10] We
"could have a map that showed where the fires are and where grazing
was done." Bush works with both sheep and goats. Sheep primarily graze and goats browse, which means more types of vegetation
get a trimming—the grassy ladder as well as shrubs and the brushy
understory of trees. She says this can "really reduce the speed . . . if a
fire did come through."[11]

Then there's the joy of working with animals: the sweetness, silli-
ness, and affection of these enchanting beings. A video shows Bush
standing calm among a veritable swarm of horned and bearded goats.
She trills and, in reply, they mmmmeh and behhh. It sounds like a
murmur that rumbles through an audience and then recedes. "You
can't tell me that's not the coolest thing ever," she says, laughing. "I
do everything in my power to create the best life for these animals, to
have the freedom to be a sheep or a goat to their utmost. They get a
new salad bowl every day."[12]

———

Beth says Holistic Management saved the ranch from heavy losses
three years before. It was a hot and dry season, and there were fires
in the area. A nearby lake went dry, something no one had ever
seen. "We ended up selling cows. Our grazing plan told us what was
coming. This allowed us to destock earlier, five [cow and calf] pair."

Ranchers often take a big financial hit when they try to ride out
a bad stretch of dry weather and then, desperate, decide to sell when
everyone else is selling and the price is rock bottom. This is the kind of

thing that puts cattle farmers out of business. Holistic Management involves making educated guesses about conditions, and adjusting the plan as a situation develops. The key is monitoring, observation, flexibility. Says Beth: "You plan in pencil. You put in pen what actually happens. The plan will not go according to plan. Try not to get too worked up over that—you're managing complexity. Over time you get a record. The better record keeping you do, the better you'll be able to learn."

Beth calls the grazing plan the "sexy part of Holistic Management." We are again around the big table and Beth is distributing graph paper, rulers, and pencils. Sexy? More like *ugh*: I'm one of those people who's phobic about paperwork, and am immediately convinced I'll mess it up. We take a hypothetical year on a hypothetical ranch based on the Lazy R, and Beth leads us through the process. "There's a little bit of math involved," she says. "You can use a calculator. It's not that bad." I never like conforming to clichés but as for this one, that women don't like math, I plead guilty. Fortunately, others are good with numbers.

What's important in planning, she says, "is not so much where the animals are as where the animals *aren't*. Everything revolves around recovery periods. Determining the length of recovery periods is probably the most important decision in the whole plan--the 'secret sauce.' The grasslands are not a tool to grow animals. The animals are a tool to grow grasslands. We are grass farmers first and foremost." Holistic planned grazing can be hard to explain, and Beth's focus on the recovery period is clarifying, shedding light on how this involves looking at the system as a whole as opposed to focusing on one isolated goal.

We work, surrounded by wide window views of the rangeland and Beth's two Appaloosa horses eating and swishing their tails. Beth gives us more to think about as we mark down lines indicating the dates in which the cattle will occupy respective paddocks. "A fence is a contract; you are telling the animal, *I will meet all your needs as long as you stay in the fence*," she says. "Plan for calving, and not when you're on vacation. Consider, where would be a nice place to be a calf? Not where it would likely be muddy."

She points out that one paddock borders a wetland reserve where waterbirds breed. "The animals would graze there after the water-fowl nests are gone," she says. It's like working out a puzzle. And before the afternoon break, we have each created our own grazing plan. Mine is a little smudgy, but not bad. Anyway, this is why they invented erasers.

The last evening we have a graduation ceremony in which Beth and Sandra express intentions for us as we take our new knowledge into the world. We walk around stations marked by faux candlelike lights, taking in each message. Then we sit and listen to Sandra tell a story in the form of a poem titled "The Bargain." It is a story about Queen of the Night, or Queenie, an especially serene and beautiful heifer who was injured at seven months at a time when Sandra was off-ranch. For months Queenie was at the edge of living and dying. "Every day she should have died, but she didn't . . . Take a bite of hay, I pleaded. Take a drink, I implored." Distraught, Sandra wrote, she "made a bargain with God" to let her recuperate, "frolic and play on the hilltop again," and have a calf of her own.

Queenie recovered and went back to playing on the hilltop with the others. Two years passed and Queenie bore a calf. Everything went well and she loved that baby but the night after the birth Queenie died. In Sandra's words, "The bargain had been fulfilled." She named the little guy "Dewey" for dewlap, the loose skin under a cow's neck, and bottle-fed him. She calls him a "sweet, big puppy-dog of a calf. He will be the farm mascot for the rest of his natural days."

Our week at the New Cowgirl Camp has dispelled some of the mystery of raising livestock in a way that restores the landscape.

But—as the story of Dewey and Queenie reminds us—not *all* of the mystery.

SEVEN

Science and Inspiration

What the Rain in Spain Foretells of Our Future

The point about the firewood is that there was practically no firewood to be had. Our miserable mountain had not even at its best much vegetation, and for months it had been ranged over by freezing militiamen, with the result that everything thicker than one's finger had long since been burnt.

—George Orwell,
Homage to Catalonia, 1946[1]

S pain has among the worst desertification in Europe, with signs that barren, arid conditions are creeping up from North Africa. That parts of Spain look like a desert is hardly news: Almería province's "badlands" area, dubbed Mini Hollywood, has served as the backdrop for numerous films, including *Indiana Jones, Lawrence of Arabia,* and Sergio Leone's spaghetti westerns. More recently, large parts of the country that were formerly green and fertile, including holiday areas along the Mediterranean coast, are contending with topsoil loss, erosion, and rivers running dry.[2]

At the same time, many people in eco-restoration circles see Spain as ripe for change, largely due to its location, relative stability, inclusion in the European Union, and engaged communities of farmers. Two initiatives in Spain—AlVelAl in Andalusia, which aims to realize the

economic potential in landscape restoration, and Camp Altiplano, the
first Ecosystem Restoration Camp (ERC), which seeks to launch a
global, grassroots movement to restore degraded lands—represent two
dynamic models to engage people in landscape restoration.

How did much of Spain's landscape become so degraded? To
understand that part of the story, I sought out Valencia-based scientist
Millán Millán, who for several decades has been applying "ecological
forensics" to the region. His research on how the environment has
changed over time offers valuable insights for how to restore ecological conditions here. While he focuses on the Mediterranean Basin,
his work has implications for many areas where rain patterns have
shifted and land is drying out.

Professor Millán Millán (the two *ll*s are pronounced as a *y*) is
a strong, often lone, voice within the scientific community, arguing that changes in land use have profound effects on regional
climate patterns. He has been observing these changes along the
Mediterranean coast for decades, often moment-to-moment as he
watches clouds advance and retreat from a small weather observatory
in his thirteenth-floor apartment. His work points to a clear connection between landscape alteration and climate, a subject generally
absent from high-level discussions about climate change, which
typically focus solely on parts per million of carbon dioxide in the
atmosphere. Millán is spoken about with great reverence by John Liu
and others. Rhamis Kent quotes his aphorisms: "Water begets water"
and "Vegetation is the midwife of precipitation."[3] How often had I
heard, "You *must* talk to Millán Millán," almost as if this were a rite
of passage in a secret club.

A meteorologist with doctorates in atmospheric physics and
industrial engineering, Millán invented the technology underlying
the metal detectors used in airports and designed something called
a correlation spectrometer (COSPEC) to measure remote atmospheric gases. The device, originally designed to track pollution from
industrial smokestacks, is used for detecting volcanic activity as well.
Millán has also, since the 1970s, helped define environmental and
climate research goals for the European Union.

Millán's recent work focuses on the loss of the daily summer storms in the Western Mediterranean and the link to changes in land-use patterns. However, loss of rain is not what's happening on the November day Tony and I arrive. From the moment we catch a taxi from the airport, it is coming down in torrents. I'm wearing a red, mountain-weather rain jacket, clutching at the hood to keep my hair from getting soaked. My boots and jeans become a lost cause when I cross a street and slosh through a puddle. It usually doesn't rain this late in the year, our B&B host tells us apologetically as he hands us the keys.

"The weather report said it was supposed to be sunny," I lament when I get Millán on the phone. He laughs and says, "Haven't you heard the phrase, *You lie like a meteorologist?*" He adds, "Forecasters never get it right in on the Mediterranean side because they don't consider the effects of the sea surface temperature."

He launches into a disquisition about the current weather pattern, from which I am able to piece together a few bits: In the trade this is referred to as a back-door cold front, when moisture come from the Mediterranean side (a Levanter); low pressure from western Portugal is helping to drive it; we happen to be sitting at a convergence of moist air masses, hence the heavy rain; and warm sea temperatures are perpetuating the system.

"It normally stops when the Mediterranean cools off," Millán says. "It's like when you blow on your cup of coffee to cool it down. After a while it cools down and you can stop blowing." After the call I hang up my damp clothes, wondering if they will ever dry.

———

Given his credentials, I am relieved to find Millán chatty and unassuming. At seventy-seven, he is energetic and knowledgeable and enthusiastic about all sorts of things, from Ladino music to mushroom foraging. He is a small man, trim in the manner of older men vigilant about their health; he had heart surgery a few years back and is careful about his diet. He apologizes that all he has in the apartment is a bag of oranges.

He tells us a bit about his family background, which, through the ages, has included landowners and rebels, converted Jews, Arabs—many of the historical strands that make up Spain. Millán grew up in Granada in an old family house on a hillside that can be seen from the Alhambra, the stunning complex of palaces and gardens that represents Moorish Spain at its peak. As a child, he and his friends used to tease the guards by throwing a ball and sneaking past the gate to retrieve it. His father's life was saved twice by the same man during the Spanish Civil War. Life in the late Franco period was bleak and constrained, but Millán got out—initially as an exchange student in the United States and then to university in Canada.

He returned to Spain in 1981, after Franco had died and the country was primed to enter the European Union, and became active in environmental research. Millán has little patience with bureaucrats and careerist academics, of which he assures us Spain has plenty. A reminder of such sorts unleashes a stream of complaints and epithets, in a blend of Spanish and English. I venture that this, the circulatory system benefits of a good rant, helps keep him fit.

In the 1990s Millán was asked by the European Union to research weather anomalies in the Mediterranean region. Basically, there were twenty years of data that deviated from accepted patterns. This launched a path of inquiry that led him to conclude: What we do to the land alters the water cycle, and this has an influence on climate. He determined that human impacts on hydrology were already contributing to extreme weather events such as prolonged droughts and intense, damaging storms. This research also gave him the conviction that once we understand the dynamics that lead to these negative scenarios, we can implement land management approaches to recover previous rainfall regimes. That is, water cycle disruptions are reversible. This body of work strongly supports the importance of ecosystem restoration.

Over the subsequent three decades, Millán participated in multiple EU scientific projects, analyzed historical data, ran simulation models, lofted balloons, and employed the latest mapping and measurement instruments to better understand evolving conditions in the Western Mediterranean and the relationship to broader trends. The research explained phenomena he was witnessing in real time, namely that coastal rain patterns were changing. He points out that like Europe as a whole, Spain's eastern region straddles two hydrological basins, the North Atlantic and the Mediterranean. The Mediterranean Basin, which most strongly influences the area around Valencia, relies for precipitation on moisture that is recycled within the catchment. He learned that human-driven impacts on the land determine both the amount of water recirculating and how that moisture is distributed.

Millán explains that in the past, predictable daily summer rains cleansed the air and provided water for farming. The storms acted as a bridge, delivering water in a season when other weather systems are at "rest." Water that evaporated from the Mediterranean Sea would fall at higher elevations some sixty to eighty kilometers (thirty-five to fifty miles) inland— the moisture climbing up the slope, as if on a staircase, via the orographic effect that lifts and then cools the air mass so that it condenses to form clouds and rain. The culmination of several land-use changes over time—deforestation, the draining of marshes for agriculture, construction on wetlands, shrouding soil with houses, asphalt, and concrete, and an increase in wildfires—has resulted in less evapotranspiration along the sea air's path.

On its own, the vapor carried from the Mediterranean was never sufficient to elicit rainfall, he says. The moisture rising from the land, emanating from moist soil and transpiring from vegetation, was the extra trigger that made it happen. And so its absence leaves a deficit. Millán has noted that in this particular environment, the concentration of ambient moisture necessary to produce rain is twenty-one grams per kilogram of air. The average water content of the summer sea breeze is fourteen grams per kilogram of air. Without that additional seven or so grams of vapor picked up over the landscape, the sea breeze cannot scrounge up enough moisture to make a storm.

The missing rain isn't the only effect of land degradation, says Millán. With less vegetation transpiring moisture, the land surface heats up. Warm, barren land repels clouds, which retreat and cause precipitation over the water. Says Millán, "If you accumulate enough water vapor, you get a storm. If not, the moisture goes back toward the sea."

Also, much of the water vapor that would have fallen as mountain showers is still hanging around. This moisture-laden air absorbs pollution and contributes to a greenhouse effect that warms the sea. He notes that the temperature of the Mediterranean Sea has increased nearly 4°C (7°F) over the last thirty years. The lack of consistent storm activity puts the catchment area in an "accumulation mode," whereby the lingering vapor and pollutants pile up in layers above the sea, creating a kind of meteorological traffic jam.

That vapor has to go somewhere, and much of it moves through advection—horizontal transfer—to where it can condense. Millán's research has tracked this fugitive moisture to precipitation events in Central Europe, which has experienced tremendous flooding in recent summers. In his words: "Basically, you're cutting a tree in Almería and getting a storm in Dusseldorf." When it does rain near the shore, it is more severe, resulting in floods, mudslides, and erosion. The new pattern has been less rain in the summer, when it's needed, and more heavy rains in the spring and autumn. It becomes a self-perpetuating cycle: Without regular summer rains the sea is less able to discharge heat, while the warmer sea surface feeds out-of-season coastal downpours.

This last bit is the cold-air-over-warm-water scenario, like the frontal system that's been drenching us since we got off the plane. It results in dramatic swings between aridity and inundation, which exacerbates desertification and all the problems associated with floods, from crop damage to water contamination.

Says Millán, "When there's development along the coast, the convected condensation level goes up." This means that a parcel of air needs to reach a higher pitch of elevation in order for the moisture to condense. "The changes predicted for fifty years from now are happening now. We have exotic species coming into the sea. Those regular rains replenished the aquifers. Now we have drought, but when the rain comes it's heavy, which leads to floods, mudslides, and soil loss."

Not a stellar situation, and one that's reflected throughout the Iberian Peninsula and is putting pressure on already struggling farmers. However, Millán believes that if we can understand the circumstances that interrupt long-standing weather configurations, we can develop tools to reinstate hydrological cycles. He believes, in other words, that it is possible to "cultivat[e] summer storms."[4] This can be done, he says, by ensuring sources of evapotranspiration (in other words, vegetation) along the sea breezes' path so that the air mass picks up the requisite moisture. Reforestation in areas that had previously provided vapor for the daily rains can, to some extent, compensate for previous damage.

Research into soil-plant-precipitation interactions can also inform where and how to design, build, and develop human habitation in order to safeguard environmental function, he says. For example, in the Western Mediterranean Basin and coastal zones like it, there should be forested areas within the catchment. Homes can be painted white to reflect, rather than absorb, heat. This way, sea breezes will not gather heat as they sail above the landscape. By creating the conditions under which the climate has functioned well in this region, Millán contends, it is possible to stabilize the hydrological cycle on the local-to-regional scale. He does add the caveat, though, that that the second feedback threshold, a massive loss of soil, cannot be crossed.[5] This is because such soil-surface properties as albedo (reflectivity), heat absorption, and moisture retention have tremendous impact.

Judiciously reinstating vegetation and permeable, non-heated surfaces in an area, one could induce precipitation sixty or eighty kilometers (thirty-five to fifty miles) upslope or downwind, depending on the natural course of moisture. This is a remarkable prospect: the possibility of undoing ecological disruption, of reconnecting what has been severed—in this instance the flow of water vapor. What Millán is describing is similar to what's been done on the Loess Plateau and what is planned for the Sinai and other desert regions, and, for that matter, the revival of grassland areas with holistically managed livestock I've witnessed in Zimbabwe, Mexico, and many places in the United States. It is another example of earth repair, and the fruit of years of scientific investigation. The more we know about how natural systems work, the better able we are to mend them when they falter.

———————

The next morning Millán picks us up in his small, agile white car. We are headed toward the continental divide, a journey that, he explains, offers clues of how human activity has led to the semi-arid, fire-prone landscape we have now. As we head inland, we are also heading back in time to witness the ways that local people, tyrants, fascists, and various marauding invaders have wreaked ecological havoc—and given Millán years of material to study.

The weather has granted us a reprieve. It is no longer raining, though the sky is opaque and the air is heavy, as if to remind us our luck could change at any moment. Periodically, Millán comments on the movement of industrial plumes or shifts in wind and estimates when, and where, the next rain will begin. His accuracy is impressive. I have never known someone so attuned to clouds.

In downtown Valencia, the streets are lined with orange trees and modest tapas bars, tall palms, and, down the hill in the Old Town, the dirty blond of historical buildings. We cross the Turia River, which after a devastating flood in 1957 was diverted to create the park that now runs through the city. While pausing at stoplights, Millán voices his fraught relationship with his homeland. "While other countries had the industrial revolution, Spain was investing in promoting the idea of Mary's immaculate conception," he says, shaking his head. He says that grievances from the Spanish Civil War still fester beneath the surface.

Beyond the city we can't quite see the Mediterranean through the haze, but Millán points out one of the few remaining marshes. The road is slowly rising in elevation. These are foothills where the Romans built roads, he says. He gestures toward a riverbed. The Romans and Carthaginians fought here, though "only the Romans knew how to navigate the river." The farms and orange fields on both sides of us benefit from soils washed down from the deforested mountains, he laments, noting the folly of basing entire industries on circumstances created by environmental degradation. How well is that likely to turn out? For the time being, however, this area, Campo de Turia, gets four crops a year.

We continue past villages along the lower slopes. To me it is picturesque—clusters of venerable-looking buildings surrounded by fields, and the occasional stone relic—but Millán says the region tells a "story of awful land management. For hundreds of years people built houses in rocky outcroppings in the mountains. They farmed in the flatlands, and then people built in in the flatlands. All of this was marshland." When intact, such marshes would absorb, release, and filter water. Wetlands, in other words, "keep the water cycle turning" and are an important source of moisture as the seaborne air moves inland, coaxing vapor along the way.[6]

Farther along, still in the midst of farmland, we pass the town of Sagunto. I can see the profile of a castle and some putty-colored stone walls. This is where Rome was taken by Carthage during the Siege of Saguntum in 219 BC, an assault that triggered the Second Punic War. Under Hannibal, "the Carthaginians would attack in the afternoon when the setting sun killed the Romans' vision," Millán tells us. By now we're past the large farms. The hills are rocky with sparse trees.

"In the 1600s and 1700s there were tremendous flash floods here," he says. "And tons of silt because forests were burned to make pastures." He seems to know the provenance of every stand of trees: We soon encounter thicker forests replanted after the Spanish Civil War, and near a turn in the Palancia River we pass a reforested area that succumbed to wildfire twenty years ago. With recurring fires, tinderbox conditions evolve. Plants that rely on fire for germination, like pines, start to dominate. "One thing people don't understand is that you can tilt the system," says Millán. "They planted pines, which are fast-growing, but there was no management."

The pines are not native, in contrast with the scrub oaks they replaced, oaks that are now being cultivated for truffles. "In the hills, you can see the history of the fires. These places have been cultivated for two thousand years. The forests were owned by the church. Trees were cut to make cathedrals. Whenever they needed money, they cut the trees. And during wars, they burn everything." The civil war was especially bad, he says. Millán's wide-ranging knowledge of this land informs his understanding of how it has been influenced by people for thousands of years, and what can be done about it now.

We're rising in altitude, now at three hundred meters (a thousand feet) above sea level, and, it seems, moving backward chronologically. We pass thousand-year-old hamlets, and the remnants of Moorish terraces, built before the late 1400s when, along with Jews, North African Muslims were expelled. ("The Moors in Valencia owned the best land," says Millán.) We pass a Roman watchtower, and a Roman footpath that runs along a river. The successive changes in landscape and architecture indicate the changes in the environment over time.

Dotted about are smallholder farms. Because water for irrigation is less dependable, and sometimes absent, people are digging deeper and deeper wells, says Millán. As a consequence, the water tables are plummeting. An old house and derelict sheep pasture emerge through a pearly mist. In this area, past where the main river was navigable, "it was probably paradise, with oranges, cherries, apples, and persimmons," he muses. "I used to love the landscape. Now I see all the problems." It is an occupational hazard, to be sure.

We climb toward the second up-rise. This, says Millán, is the second stage for storms. If the sea breeze, moving inland, lacks the moisture for rain, it lifts as if ascending through a chimney toward the damper, cooler air more conducive to precipitation. He notes that several houses have archways. "This suggests the land was cultivated by expelled Moors in the late 1400s and 1500s." Approaching a thousand meters (thirty-three hundred feet) we see abandoned water channels near a onetime Roman path. Ahead there is fog: moisture-laden clouds. It's as though we're chasing the mist.

We continue toward the fog, flanked at once by oak plantations and swathes of land without trees. "Cut trees and you get drylands," he says. "It floods and you think you have water. This village now has no water. What they had was excess water from the channel." Millán says we're at the edge of the Mediterranean water catchment—and therefore of where the rainstorms pick up their haul. To the right he indicates another small village. "There were natural springs here. Now there is no water. They bring it in in trucks. The politicians say nothing."

At 1,223 meters (4,012 feet) Millán says we have crossed the continental divide. With no road marker or fanfare we find ourselves on the

Atlantic side—at the hydrological frontier. We're in the province of Aragon, near its capital, Teruel, where George Orwell was wounded during the Spanish Civil War. Nearby are the headwaters of the Tagus River, the longest in the Iberian Peninsula, flowing toward Lisbon in Portugal. "We've hit warmer air. It's no longer moisture from the Mediterranean," he says. It's one thing to know that the region is divided into two maritime environments, and another to see, smell, and feel it.

There are more pine plantations. And deserted fields, where Franco had promoted irrigation for impoverished rural people, schemes that couldn't hold even for a single generation. A new "airport" (Millán: "Where companies dump old airplanes to be refurbished") and a new paper mill ("Another way to deplete the trees"). We stop at a hotel, Parador de Alcañiz, for an early lunch (early for Spain, that is, where the midday meal, the main meal of the day, is between two and four). It is a castle and monastery that dates back to the twelfth century. I stretch my legs and enjoy the mild, fresh air. The sky is blue. We seem to have outrun the clouds and entered a different world from the rain-drenched Valencia coast.

———————

Our final morning in Valencia, we meet at Millán's place for oranges and coffee and he shows us graphs and images from his research. In our discussions he'd said the association between devegetation and a decrease in rain had been known since the 1970s. "Many of the dynamics we are seeing now were alerted back then," he says, contending that land-use impacts on climate should have been a focus of ongoing research. Instead, he says, scientists grew enamored with greenhouse gases and CO_2 modeling. This became the go-to topic for research and academic status at the expense of other factors.

Millán refers to two prominent reports published by the Massachusetts Institute of Technology Press, *Man's Impact on the Global Environment* and *Inadvertent Climate Modification*, published in 1970 and 1971 respectively. He says his mentor, Canadian meteorologist and mathematician R. E. (Ted) Munn, wrote a chapter on the "climatic effect of man-made surface changes . . . One day he brought a manuscript and said to me, 'Can you please check over chapter 7?'" he recalls. "I was one of the first to read that chapter."

I track down the books in the science library at Williams College—amazingly enough, the very place where the scientific meetings that led to the first book had taken place back in July 1970. Munn's chapter is in the second book. It begins: "There is no doubt that whenever man changes the landscape he modifies the microclimate."[7] The piece considers everything from "urban pollution plumes" to the way that paved roads affect the heat budget. It is affecting, and more than a little disheartening, to browse these volumes from the vantage point of 2019. One section of the first book concludes, "It seems obvious that before the end of the century we must accomplish basic changes in our relations with ourselves and with nature."[8] To be clear, the century the authors referred to is the twentieth.

Land-use climate effects are something "nobody wants to hear about," says Millán. "They represent a more uncomfortable truth. Greenhouse gases are very convenient: You can talk about it a lot and do nothing about it. Land use you can do something about. But then we're on the edge of something very uncomfortable. Try telling people they can't develop on the coast because it can affect conditions fifty or sixty miles downwind." He contends that much of the weird weather we're seeing across the globe—floods, droughts, and heat waves—is the result of local impacts to the water cycle as a consequence of what's been done to the land. The emphasis solely on carbon emissions has blinded people to what's happening right in front of them, he says. "People aren't looking at local-to-regional effects."

He also expresses concern about current science's reliance on models: "Models do not replace observed phenomena. You have to use models judiciously, based on real data," or, he says, in conjunction with other research methods. "With modeling there's often the notion that there is a clear answer. Or that all one needs is a better model. But it's an iterative process—not one thing."

Millán shows us temperature profiles taken by an instrumented aircraft and photographs of inversion layers over the Mediterranean Sea: distinct bands of water vapor suspended in the atmosphere.

"That's warm air on top of cold air. They don't like to mix," he says. The challenge, he says, is that these strata may not show up in the models: "The models average, or the layers are smoothed out in the presentation. It is the failure of the numerical system. Even the best resolution cannot incorporate this."

That climate and weather models may not be as comprehensive—as infallible—as we think could be upsetting or reassuring, depending on how you look at it. But perhaps it doesn't matter. The point is that we need to be alert to what's happening in the natural systems we engage with, whether it's the Mediterranean Sea or the Arabian desert or the charred plains of eastern Washington. For ultimately, ecological function—and by extension climate regulation—is about place.

The Business Case for Healing Land

Tony and I encounter our next big rain two days later in Chirivel, a village in Almería province that is surrounded by mountains and boasts relics from Iberian, Arab, and Roman times, including a statue of Dionysus from the second century. We're at the Torregrosa family's farm El Chapparal for an outdoor lunch. The farmers are members of the AlVelAl Association, a local initiative working in collaboration with the Netherlands-based company Commonland. It's a jolly event to celebrate one farmer brother's birthday, introduce AlVelAl board members to the community, and showcase AlVelAl members' wares.

There must be fifty of us milling about, and enough food for two hundred. Long tables laid end-to-end are piled with local food products: jars of honey; bowls of almonds; thick wedges of spelt bread made from local grains; robust red wine (labeled and label-less); the ubiquitous blond Alhambra beer from Granada; and a dizzying array of cured sausages and sliced ham from the family's organic farm.

The property is on a plateau, and the wind is whipping up. There are some sturdy oak trees scattered around, but not enough to offer much shelter. Large clouds glide past, alternately baring and eclipsing a sharp blue sky. I refill my glass of red wine for the warmth. A round of speeches begins, extolling the agricultural renaissance that's stirring here in eastern Andalusia. Suddenly the sky breaks open and the rain crashes down. Some moments of chaos ensue as people scatter and rush to find room in the vans waiting on the gravel. A few remain at the grill tending large pork loins that have yet to be served, looking wet and forlorn.

El Chapparal, which specializes in raising and processing black Iberian pigs for high-quality meat and to regenerate the landscape, is part of an ambitious regional project across seventy-six municipalities in the high plains of southeast Spain. The venture begins in the Netherlands with Commonland and its quest to create opportunities to invest in land restoration. Which often means significant time and financial commitments to seriously degraded landscapes such as this, the Altiplano region of Spain, where conditions have deteriorated so that farmers are abandoning land and young people in search of a future have gone elsewhere.

Commonland's CEO is Willem Ferwerda, a Dutch ecologist specializing in tropical habitats. For more than a decade he was the executive director of the Netherlands office of IUCN. When I met Willem at the 2017 Caux Dialogue, he told me that after years of developing and funding more than a thousand well-intentioned, scientifically impeccable conservation projects while watching global ecological conditions deteriorate, he'd reluctantly concluded that the current conservation model wasn't working. In 2009 Willem heard John Liu speak about the Loess Plateau Watershed Rehabilitation Project at the Tällberg Forum in Sweden. As the story goes, he approached John and said to him, "We are going to be working together for the rest of our lives." John has an ongoing role at Commonland as ecosystem ambassador.

Willem devoted the next few years to the question: *How might knowledge and resources from the business world be integrated into large-scale ecological restoration?* Clearly, the way business is conventionally done is less than kind to the environment. This is largely due to a design flaw in our global economic system: The value of nature is zero. It's that old trick of externalizing negative outcomes so that someone other than the company who caused the problem has to deal with it. However, investors and others across the business world increasingly see this as untenable. Many are exploring models for a regenerative economy, whereby the production of goods bring ecological benefits. A few examples include: the Landscape Finance Lab, through the World Wide Fund for Nature; the Savory Institute's Land to Market program, which verifies regenerative sourcing for meat, dairy, leather, and wool; and Regen Network, which uses blockchain technology to create incentives to improving ecosystem function.[9]

While not naive about the nature and priorities of commerce, Willem believed a business case could be made for land regeneration—and that if done right, this would not merely mobilize capital but also unleash innovation. Early on he determined that unlike many business enterprises, which accede to the impatience of the market, the time line for investment earnings would need to be twenty years; ecological restoration is not so easily wedged into quarterly reports. The timescale and the breadth of the endeavor would require a shared commitment on the part of all stakeholders: producers, investors, and communities. Commonland launched in 2013.

The model that Willem and colleagues developed combines two basic concepts. One is classifying areas according to zone: natural, which prioritizes restored ecological function (to "create stepping-stones for flora and fauna," as one manager put it); economic (sustainable business ventures and value-added production); and a combined zone (regenerative agriculture). This reflects lessons from the Loess Plateau, where the team found that preserving a section for ecological integrity improved the productivity of nearby farmed areas.

The hallmark of Commonland is 4 Returns: Financial Capital (without which one can't play in the business realm); Natural Capital (the venture must be ecologically regenerative rather than extractive); Social Capital (giving back to the community and enhancing local well-being); and Inspiration (offering hope for the future and a sense of purpose). This last component is not to be taken for granted, particularly in areas like the Altiplano and counterparts across the globe that have been in decline for generations.

Commonland has developed four projects in four different places to test their 4 Returns framework:

One, in partnership with local organizations Living Lands and Grounded, is in remote Baviaanskloof (Dutch for "Valley of Baboons") in the South African Karoo. Two centuries of poor livestock management—sheep and angora goats for wool—have degraded the landscape, which includes the catchment area supplying most of Port Elizabeth's water. Through working with *spekboom*, an indigenous succulent that aids in restoring land function, and cultivating perennial plants for essential oil products, the project helps farmers and local workers reboot the soil, regain water security, and provide employment and training in a place where most have few options. The effort influences five hundred thousand hectares, or five thousand square kilometers (two thousand square miles): larger than the state of Rhode Island.

In Western Australia, Wide Open Agriculture is bringing diverse, ecological farming and sustainable water management to the struggling Wheatbelt region, an area that has lately trended toward "bigger farms, bigger machinery and less community." The Wheatbelt has been losing population at a rate of 2 percent a year, leading to closing of health facilities and grade schools. One project focus is reaching out to "new Australians," recent immigrants including asylum seekers and other refugees. The company raised $5 million in its July 2018 initial public offering and is listed on the Australian Stock Exchange (ASX). It recently launched its own brand, Dirty Clean Food. Wide Open Agriculture's land spans three hundred thousand hectares (three thousand square kilometers or twelve hundred square miles), more than three times the area of New York's five boroughs.

Closer to home is Wij.land, which seeks to restore peat meadows: that quintessential Dutch landscape complete with windmills, flowers, and picturesque cows. These peat-y fields are rich in stable carbon and important for water flow. In recent decades, however, they have been intensively farmed and drained, resulting in subsidence and soil degradation; they have become "green deserts" that look pretty but lack biodiversity. Wij.land is primarily focused on regenerative dairy through planned grazing. This involves 125,000 hectares (three hundred thousand acres): more than five times the size of Amsterdam.

According to my arithmetic, the land Commonland is influencing exceeds three New York Cities, five Amsterdams, and a Rhode Island. And this is before counting the largest: here in Spain. (Projects in four more landscapes—Haiti, Kenya, India, and Zambia—are also under way in collaboration with local NGO partnerships.)

As they assembled their portfolio, the Commonland developers wanted to work on a project in the Mediterranean Basin. According to Commonland landscape manager Erica ten Broeke, in 2014 the team took a scouting trip to Spain, driving around the countryside and meeting with local people. In the Andalusian high plains they found farmers receptive to their ideas and an agricultural heritage that still resonated, including long traditions of water management. The Commonland team put together a series of workshops in which they developed a shared vision using the participatory group process Theory U.[10]

In April 2015 Asociación AlVelAl was established. The name incorporates the three adjoining regions that comprised the initial focus: the *Al*tiplano (plateau) of Granada; Los *Vél*ez; and *Al*to Almanzora. (The project area has since expanded to include northeast Murcia and Guadix.) "In the first few years Commonland and AlVelAl managed jointly," Erica says. "Now AlVelAl is taking the lead." She says 120 farmers are now involved with AlVelAl, with 63 actively implementing regenerative practices and more making the shift. Many others in the region are involved through business/entrepreneurship, community organizations, and research. In addition to direct support and training for members, now numbering about 250, Commonland/AlVelAl regularly hosts public educational sessions and "agri-cafés," and uses local media to build and maintain community.

The total area is a million hectares (ten thousand square kilometers, or almost four thousand square miles), the size equivalent of Lebanon. After having gone on long drives between AlVelAl farms and member enterprises, I can attest that it covers a lot of ground. The rolling plateau is big, mostly empty land. As our son Brendan, who joined us on our last day in Valencia, said, looking out the car window, "It's like a flyover state." Since he himself grew up in a rural state, presumably he meant this ironically. Still, this is precisely the perception—and self-image—that the Commonland partnership has vowed to change.

———————

Our base for the visit is Vélez Blanco, a sixteenth-century village of whitewashed buildings and narrow roads winding up the hillside: a vanilla-frosted layer cake of a town with a Renaissance castle on top. The castle's marble patio has been removed to the Metropolitan Museum of Art in New York, leaving the otherwise impressive stone fortress with one blank wall. The town sits at the approximate center of AlVelAl's terrain and ten kilometers (six miles) north of Vélez Rubio, a larger town that, as its name suggests, has ruddier-colored buildings and rooftops. It also is at the edge of the Sierra María–Los Vélez National Park, which has pine and Mediterranean woodlands.

We're staying at the sprawling, rustic El Palacil, the kind of hotel that entrusts its guests with actual physical keys. Once a water mill, it is a family operation with all of the reception and food service handled by the attentive, competent, and—word has it—soon-to-retire Isabel Rosa. It is near enough to the center that you can walk to the water museum, which depicts the history of the water network built by the Spanish Moors. The system, marked by complex sharing and management arrangements, collected water from mountain springs and distributed it to townspeople and farms. Nearby is the Cañon de la Novia (Fountain of the Bride), where many people still get their drinking water. You can stroll along the channels that have served the community for more than a thousand years. Vélez Blanco's hydro-history is everywhere in town.

· The next morning we meet as a group—joining me are three representatives of the Portuguese company Sementes Vivas, or "Living Seeds," and two investors based in Spain—for an overview of the project. AlVelAl vice coordinator Catalina Casanova stresses the social aspects, specifically the attempt to support farmers.

"Farmers have been seen as less intelligent than city people," she says. "Maybe once a year you need to see a doctor, but three times a

day you need a farmer. We believe in the dignity of the people working in this landscape. We want to give people an alternative to public
aid." She says it's important that the premium for added value would
go to the producers, not a middleman.

The Altiplano project area is almond country, with a hundred thousand hectares (250,000 acres) in rain-fed almond production, almost
half of them organic. (Eighty percent of global almond production
is in California, and almost all depends on irrigation.) It was during
the Moorish period that almonds were brought into the region as an
important crop, Casanova says. Recent years have brought an almond
"green gold rush," in part because the price of almonds in California
went up as access to water has become more difficult. Keen to cash
in, farmers in Andalusia turned to monoculture almond plantations.
And now the price of almonds has fallen. Monoculture nut farms pose
additional costs, she says, as topsoil is lost under poor management.

Commonland/AlVelAl's response is the almond production
ecosystem *Almendrehesa*, a play on the word *dehesa*, a type of southern
Iberian landscape that integrates woodlands, farming, and pastoralism. Conservation biologist Astrid Vargas, who was instrumental in
reintroducing the Iberian lynx, helped design the model. In a video
Vargas explains why the Almendrehesa makes ecological and business sense. Indigenous Segureño lambs aerate and fertilize the soil,
and their grazing trims the vegetation so that it doesn't grow, dry
out, and risk becoming tinder for summer fires. At the same time,
this eases the grazing pressure in the natural zones. Native aromatic
plants like lavender, thyme, and rosemary are planted between trees,
and become the base for organic essential oil enterprises. Their planting also prevents erosion and provides habitat for pollinators. The
presence of pollinators, particularly endemic bees, increases almond
tree productivity.[11] By contrast, California almonds rely on pollinators
brought in from elsewhere via the bee-brokerage market. According
to Tom Philpott of *Mother Jones*, some 1.7 million bee colonies per
year are shipped to almond groves in the Golden State.[12]

When sheep graze among stands of almond trees, this creates
"a mosaic instead of a monoculture," says Casanova, noting that

silvopasture can also be applied to olives—Andalusia is the world's top olive-oil-producing region—pistachios, and other crops. "If you have sheep, you have ground cover," she says. "Farmers say that sheep will be competition for the almonds, but this raises production and biodiversity and lowers the cost of fertilizer." The presence of animals also, she says, beautifies the landscape.

The new La Almendrehesa brand has secured contracts in Germany and the United Kingdom for its flagship product, Pepita de Oro (Golden Nugget): lightly roasted almonds that come in a compostable package. The almonds are different from California varieties. The flavor is subtler and somehow cleaner, with slightly floral tones. Says Casanova: "Almonds are little ambassadors for our region."

On drives through the Altiplano, we see almond groves everywhere: geometric configurations of tree rows set at regular intervals in the red, high-mineral soil. Depleted though the land may be, it is pretty country. With the volatile weather, the light is dramatic, with sunbeams falling on green fields—a rainbow seen from multiple angles as the roads curves around. "Every time you cross a mountain range, the landscape changes," Erica says en route.

We visit two AlVelAl demonstration farms. The first, run by women, is experimenting with combinations of crops, including almonds, grains, and livestock. All farmers here have to contend with soil that's low in organic matter, skeletal soil that quickly dries out and easily erodes. With sheep and cover crops, says farmer and AlVelAl board member Santiaga (Santi) Sánchez, "the soil is building like a sponge for rains. This gives more life to the almond trees, and we have nitrogen fixation with legumes." The almond trees in these fields were planted in 2015. The neighbors' trees, just over the property line, were put in in 2012. "In one year they will be the same size," says Santi.

Though she has been nationally recognized for farming innovation, like many local regenerative farmers Santi often finds that her work is met with suspicion. "I'm trying to convince by example," she says, noting that her neighbor thinks she's foolish for spending time and money on green cover. "It's a slow mental shift. People are conservative and have to see the example. People will start thinking." She points out the tumbleweed, an indicator of bad soils, on the neighbor's grounds. She says: "They are plowing five times a year."

But for the mist over the mountains, the sky is now clear. The sensation of sun on my back after several days of rain and clouds is making me sleepy. I nibble on a sprig of arugula, which is growing in dense tufts everywhere, then crouch down and grab a fistful of cool, crumbly soil. Santi draws our attention to sections of tall esparto grass, which retains soil organic matter and runs deep roots. Its fibers have long been used for baskets and ropes.

The next visit brings us back to Chirivel, where we find Manuel Martínez Egea's farm. Manolo, as he is known, is president of the Sabina Milenaria Cooperative for ecologically produced cereals, almonds, and wine. I recognize the name Milenaria from the previous day's labeled wine, the full-bodied, spicy red that fortified me against the wind. The farm devotes one and a half hectares (four acres) to grapes. Alas, says Manolo, "there is no harvest this year because of hail." I recall that Millán mentioned that at his place in the mountains above Valencia, hail left his beloved quinces full of holes.

Manolo says this area is "good for almond trees but not necessarily for almonds," as there is a risk of frost until June. He is planting pistachios to diversify his crops. "They are very complicated to harvest. And one in eight or nine must be a male, and those don't produce," he says. "I like a challenge."

Three of the four project locations—Baviaanskloof, the Australian Wheatbelt, and Spain's Altiplano—are sites where ancient indigenous art, particularly rock art, has been found. In South Africa and Australia these enigmatic carvings and paintings, clues to the physical and spiritual lives of our oldest human relatives, may date from tens of thousands of years ago. Here in Vélez Blanco, at the Cueva de los Letreros (Cave of the Signs) UNESCO World Heritage Site, the images are relatively fresh, around eight thousand years old.

After we've had a chance to dry out from our midday rain shower, we go to see the rock art. Dietmar Roth of AlVelAl leads us, as he has the key: literally, since one needs a special key to unlock the gate to the path to the rock art, and metaphorically, as Roth is full of knowledge about the region's geological and anthropological past. Originally from Germany, Roth has lived in Spain for nearly three decades. He is the deputy mayor of Vélez Blanco, and his embrace of the town is such that he's considered its "adopted son." In his role he acts as an ambassador, ensuring that the town's charms are appreciated and tourist attractions kept up to snuff. He is also a historian and has written about Vélez Blanco's water-sharing tradition, the social counterpart to the canals and reservoirs that weave through the settlements, which he says dates back to prehistoric times.

We follow Dietmar up the steps and clamber to where the rock art lives. It's a shallow cave, more of a shelter or hollow. The works we've come to see are red markings, some dark ocher and some tilted toward brown, like conté crayon: iron oxide pigments on limestone. The pictures are hard to decipher, so Dietmar has brought along diagrams. He says the art is faded not just from time, but also from visitors rubbing the stones and pouring liquids like Coca-Cola to change the hue. This was prior to the locked gate regime.

Our cave artists were pastoralists and farmers who cultivated grains, says Dietmar. They were living during the early phases of agriculture, when the relationships among stars, sun, and water took on new meaning. One of the more legible images, called *El Brujo*, or "The Wizard," has been interpreted as a shaman or sorcerer with sickles in his hands and curved horns or antlers sprouting from his head. Others portray deer and goats, flowers, or flowing water, which, he says, may suggest a spiritual journey.

Another design is the indalo man, a simple stick figure with outstretched arms bearing up the half circle of a rounded bow or, as some suggest, a rainbow. It is a spare, elegant schematic, reminiscent of a contemporary logo. The indalo has become the emblem of Almería province. You find indalo keychains and pendants and teacups around town and recognize its silhouette on shop signs and within the metal grating of windows and doorways. It is considered a sign of good luck.

We're looking east toward the area's tallest mountain, named La Muela, or "the tooth," for its appearance. The mountain has lost most of its vegetation and, as a consequence, its soil. Much of the surface, particularly at the crown, is withered down to bare rock. Although not visible from our position, a major restoration project is under way. This is a collaboration with Ecosia, a web search engine that uses advertising revenue to plant trees where their presence will yield the most benefit. The company, a B Corp based in Berlin, has projects in Indonesia, Madagascar, Morocco, Peru, Uganda, Brazil, and many other places. Ecosia has donated fifty thousand trees—Aleppo pine, juniper, black hawthorn, and oak—and is working with AlVelAl to create small check dams made from local materials to retain water and develop habitat for birds, pollinators, and other wildlife. Were it not so muddy due to the rain, Dietmar says, we would have gone to the planting site.

"What is the contribution of Almería to human culture?" he asks rhetorically. Like most places, its history reflects both achievement and defeat. We have the artists whose rock paintings speak to us across time and the culture of innovative waterworks, and at the same time deforestation, erosion, and today's poverty and despair. "We are planning to use 'land art' to restore the zone with aromatic plants on a big scale." He says the gist of the project is to re-green the skirt of the mountain—the lower third—with herbs like thyme, rosemary, and lavender arranged in the shape of the indalo. Our Neolithic forebears drew the indalo with iron oxide. Today the community is painting with plants.

"We are starting the aromatics planting next week," Dietmar continues. He says they will be working with an organization called Apafa devoted to people with special needs, giving people often excluded from civic life the chance to contribute. "We hope to cultivate a lot of pictures on the land as 'living sculptures.' Harvesting the crop can be a tourist activity"—which can provide income for Apafa. In winter the land art will be green and purple. In summer the palette shifts to white and yellow. The landscape vision is inspired by the Nazca Lines in Peru. Like these ancient drawings, it will best be viewed from high spots, such as the castle at Vélez Blanco.

The sky is broad and the clouds are dense and in motion. Perched on the rocky ledge, mere feet from ancient cave drawings, I muse on the elasticity of time: Our conversation flits from the Neolithic era to the present and to aspirations freshly dreamed. I feel the solid stone surface beneath me. Dietmar conjures yet more plans for La Muela: gardening workshops for children; archery tournaments; a national convention devoted to native esparto grasses—all on land, he says, that has not been in productive use for fifty years.

We are in the return-of-inspiration realm—that intangible yet unmistakable energetic lift. It brings me back to the consensus work in New Mexico, when we shifted from despair to best possible outcome thinking, that sense of *Wouldn't it be cool if*. . . . The Commonland-AlVelAl partnership has put that question out there: Wouldn't it be cool if instead of a blight on the landscape, La Muela was a center for community action and inclusion with the talismanic indalo emblazoned on its once-barren flank?

Catalina had spoken earlier of other inspiration category plans. There had been an event with classical guitarists playing traditional music, highlighting Spain's cultural legacy in this landscape. AlVelAl is collaborating with the U.K. social enterprise Sound Matters, which specializes in sound education and research and creates place-based sonic journeys. There will be a "sensorial path," calling attention to the aromas of various plants, and activities like tool carving to connect visitors with the prehistoric past. As Catalina says, "We take a broad view of culture."

I start to grasp the wisdom of the 4 Returns, particularly that final form of remuneration: inspiration. In an ordinary business-type business this stuff would never fly, since it is hard to reconcile with the bottom line. Yet whose idea is it that one measure represents an entire enterprise? Commonland sees four bottom lines. This makes sense to me, because businesses and other institutions are infrastructure for our daily lives and none of us is bound by one dimension.

Why should making a living be in one box and culture, aesthetics, and pride of place in a separate one? Sure, many companies have arts sponsorships, service days, and other concessions to our humanity. But here, what gives life meaning and coherence, that which kindles a spark, is built into the business model. A venture that reaps cash while the community languishes and the surrounding environment continues to degrade would not, in 4 Returns terms, be a success. In conventional Wall Street investment-strategy thinking, what takes place beyond the balance sheet is immaterial. The truth is, we are what we measure—or at least our actions are largely determined by how we gauge success. What if environmental healing, social engagement, and a commitment to the future governed our companies and institutions, and therefore our work lives?

There is no natural law that says profit must supersede other types of reward. To paraphrase E. F. Schumacher, author of *Small Is Beautiful*, we need an "economics as if people and the planet matter."[13] Many in the fields of new economics, community wealth, and the commons are exploring new models, but we're not yet there. Currently we're tethered to what we might think of as the legacy economy, defined by metrics like the gross domestic product that consider neither human nor environmental well-being.

In the essay "The Problem of Production," Schumacher remarks that given modern society's disconnection with nature, "we are inclined to treat as valueless everything we have not made ourselves." In consequence, he writes, we have confused "income items" and "capital," and have been extracting for profit irreplaceable gifts of nature that we should be preserving as capital. He offers the example of fossil fuels. If we regarded fuel held in the earth as capital, we would organize economic activity around conserving it. Because fossil fuels are, instead, viewed as sources of income, we have been heedlessly liquidating this invaluable asset. Schumacher wrote this, by the way, in 1972![14]

This kind of shift in thinking about resources, from available-for-use to capital-to-be-preserved, explains how Neal Spackman was able to grow trees in the Saudi Arabian desert. He saw captured water as capital deposits, and the team only "spent" what they could without drawing down the account. On its designated landscapes, Commonland is modeling how to build natural capital (ecological function) and social capital (revived communities) while generating income (regenerative enterprise). Here in Spain, they are trading in almonds, wine, honey, and other agricultural products.

The Altiplano project is stirring interest among younger people. I meet Belén Sánchez Martínez, twenty-eight, who studied at the University in Granada, where her final environmental science project was recognized as the best student work in Andalusia in 2015. She grew up in Oria, a small village about ten minutes from Chirivel, and is thrilled to find alternatives to simply leaving home. "Young people need to study elsewhere and then they leave the territory," she says. "AlVelAl brings the possibility of doing business here. Ten years ago it was not possible to make a life here as a young person. Now it's attracting young people. Friends who left for other cities are passionate about this project because they can see what's possible."

I chat with other young people, including Leo Pérez, also in his late twenties. He has a degree in law and economics and had been living in Granada and running a winery. He has also returned to the area to work with AlVelAl. Paco, one of the Torregrosa brothers hosting the gathering, was living in Seville and has come home to raise livestock on the farm. Augustin, twenty-three, from Belgium, came to the area to help develop a composting program. "Today only 2 to 5 percent of the population are farmers," he says, noting that this leaves 95 percent disconnected from what happens on the land.

Belén says her roots are in the community, and she wants to participate in its revival. She says of AlVelAl, "It's only three years old, but the results are big."

Camping for Change

Rehabilitating a landscape clearly benefits the immediate area. Millán Millán's research suggests that at a certain scale, ecological improvements can have an impact on larger, regional weather systems. This understanding underlies such ambitious efforts as Neal Spackman's Al Baydha project and The Weather Makers' work in the Sinai. Throughout these chapters we've seen examples of how people are initiating ecosystem restoration: large-scale government programs; community collaborations; experimental pilot projects; a focus on women; and regenerative business ventures. The organization Ecosystem Restoration Camps is looking to create a global movement of earth restorers through dedicated communities where people learn skills and become able to train others. The first one, Camp Altiplano, launched in Spain's Murcia province in 2017.

The concept was hatched during a conversation between John Liu and Rhamis Kent at the 2015 Caux Dialogue on Land and Security. For several years John had advocated for eco-restoration vocational centers. People continually asked him how *they* could help restore the earth; there had to be a vehicle to train people in restoration practice and principles so they can train others. The notion of camping—sleeping outdoors, stepping away from daily routines and expendable comforts, songs around the bonfire—caught John's imagination, and soon nearly every speech or conversation included the exhortation: "Let's go camping!" He set up a Facebook group in 2016, which to his surprise quickly had thousands of members. The page prompted discussions on everything from democratic governance and decision making to graywater plumbing and the relative benefits of tents versus yurts. People are encouraged to become supporting members by donating ten euros a month. (I have been a supporting member since the outset.)

In mid-2016 Alfonso Chico de Guzmán, a young farmer member of AlVelAl, offered to lend five hectares (twelve acres) of land for the first camp. The movement now had a home: La Junquera, a small centuries-old hamlet that had been virtually deserted. A foundation was established in the Netherlands and a website, bearing the rousing tagline "Together We Can Restore the Earth," assembled. The first group of a dozen-plus volunteers, hailing from several continents, arrived in the spring of 2017.

La Junquera is about fifty minutes from Vélez Blanco by car, and on Thanksgiving Day Tony, Brendan, and I make the trip. We pass the edge of the national park and see snow on the higher mountains. The highway has narrowed to one lane and I look left for the road going in the opposite direction, only to realize there is none, and that this skinny bit of tarmac is meant to accommodate two-way traffic. Sure enough, vehicles start heading toward us, barely skirting the chrome surface of our rental car.

This is still AlVelAl territory, but I can see that land on the Murcia side is in worse shape than the Granada side where we'd spent more time. It is the same reddish, high-mineral soil, but the fields are bare and muddy. There are fewer trees, and those look tired and wispy. We see lone, crumbling stone houses and villages, little more than clusters of derelict buildings and walls.

Finally, we catch sight of a solar array we were told to look out for. It isn't immediately obvious which is the volunteer house, and we have a brief, somewhat surreal encounter with a farmer slaughtering a sheep and throwing scraps to his dog before we find ourselves in a homey kitchen of a whitewashed farmhouse identical to others we'd walked by. Rachel Robson, my contact, greets us. "This is the base, where we have the washing machine and the WiFi," she says, and offers to show us around. She has fine brown hair and clear skin, with a crisp yet friendly manner. "We renovated the building last year. It's cozy. People are staying here, several to a room, dorm-style." There is indeed that college-dormitory air, the sense of sheets tossed aside in a rush to class—or, in this case, to the kitchen garden or compost pile. Now that it's cold outside, all volunteers but one have moved from outdoor yurts and tents into the house.

The actual camp is a few kilometers away and we've got some time, so we sit over tea. Rachel says it's now planting season, which is a challenge because the limited winter accommodations means there are fewer people around to do the work. Like Al Baydha in Saudi Arabia, they plant when it rains, at which point it can be a rush to get everything in the ground. "Water is definitely a challenge," she says. It had been a dry spring, and the size of the plants had them worried in July. Then came August with torrential downpours.

Rachel explains that the spot where we are—we can call it downtown La Junquera—consists of two farms, Alfonso's and his neighbor's. The camp has an outdoor kitchen shelter, the result of a recent group effort, as well as structures for compost and a poly-tunnel for vegetables. Other construction projects are under way, including restoring a stone round-house. "Because that is agricultural land, we can't build any permanent structures," she explains. "But we can build on existing foundations." Clearly a lot of effort is going toward setting up basic infrastructure.

I ask Rachel how they obtain provisions, and she says by now every-thing from the kitchen garden has been harvested. "There's a market every Tuesday in Topares," she says, referring to a nearby village. "We try to go when we can, and supplement with what we grow." She says the locals view the newcomers "with a mixture of curiosity, support, bemusement, and indifference. I found out in Topares that they knew us as 'the wild ones of La Junquera.'" Campers have managed to coax several crops from the sere Andalusian soil, including corn, eggplants, tomatoes, zucchini, pumpkins, melons, potatoes, and broccoli. A "dream board" on the far wall lists some extracurricular ideas for inspiration: mushroom cultivation; kimchi making; Spanish lessons; and cobhouse building. Rachel smiles. "My favorite is 'dress-down Fridays,'" she jokes.

Jonathan Church, another volunteer from the U.K., trundles down the steps, and the two share their stories of what brought them to this corner of Spain. Rachel, twenty-nine, has a background in project management and communication in local government. Her studies focused on social change/social movements and the environment. She'd been working in London for the local authority, tasked with promoting recycling and improving recycling rates. "My time outside of work, with grassroots organizations doing practical things to improve the environ-ment, became more important," she says. This included opportunities to experience nature. "Being in a natural park, suddenly surrounded by flowing water and trees: I'd physically and emotionally respond to that."

Rachel's off-hours pursuits included studying permaculture, taking bike tours of ecovillages, and getting involved with Transition Town Brixton. Transition Towns is a global movement to facilitate a positive, creative adaptation to peak oil and climate change on a community level. Transition Town Brixton has several initiatives, including a complementary currency, the Brixton Pound. "I became sick of sitting behind a desk working in an institution," she says. In spring 2018 she went on a cycling trip around Europe to learn more about grassroots environmental efforts. "This was the first project I came to. I arrived in May. I expected to stay two weeks." She is now the project and communications coordinator.

Though Camp Altiplano is only five hectares, she says, "There's a sense it has a bigger impact because of the networks it's connected to." There are usually about twelve campers on-site. In addition to working on the camp, the group works with AlVelAl farmers to implement restoration plans. One funding goal is to provide stipends for volunteers, the lack of which has been a barrier for those without income or family backing, particularly for people in Spain. Rachel and Jonathan are uncomfortably aware of how their middle-class background makes it possible to volunteer and learn at the camp. There's constant interest from journalists and others curious about the program, she says. "We've had more than eighty volunteers, and more applications than we can accommodate."

Jonathan has been at Camp Altiplano since September 2017, which by ERC standards makes him an old-timer. Also from the U.K., he is tall and thoughtful with thick, unruly hair flecked with sunny blond. It was, he says, a "moment of flux": He was working on film locations and landscaping and "looking for a way to get out of London." He is a friend of ERC business development manager Ashleigh Brown and thought, "Let me see what my friend Ash's work is all about. I realized all my work was serving the needs of two wealthy people." He wanted his work to have more meaning, and a broader impact.

"I didn't really know what I was looking for until I found it," he says. He saw in this landscape a kind of twenty-first-century tabula rasa: a blank slate full of possibility. "I thought, *We'll restore this to the lush area it can be*. I saw that part of our work was to figure out how to do that. Last year, especially, we needed to ask what ecosystem restoration means in this particular context. We've had to put more effort into organization than we previously thought. That has meant intense, intense learning, figuring out what works." He is now in charge of restoration project planning, and recently completed a report articulating strategy for the next three years.

Circumstances have often demanded improvisation, says Jonathan. The past summer they sought to implement managed animal impact. With the advice of a holistic grazing expert in the area, they devised a plan based on a density of five square meters of grass per sheep per day. The team came to an agreement with a local shepherd—perhaps the one busy carving up a carcass—and then, due to a sudden change, the sheep herder said they had to take all one thousand sheep right away or nothing.

"We had all thousand on the property for one and a half days," he recalls. "This basically meant my being a human electric fence. Not ideal, but I learned through that. I'm a vegetarian and always assumed there would be a way to do agriculture without animals." After seeing the impact on the soil, he says, "Now I'm not so sure." The team is pursuing Elaine Ingham's soil-food-web approach since a lack of plant-available nutrients is a challenge. "Getting the soil biology right is a big focus of what we're trying to do," he says.

In this region the "elephant in the room," he says, is that in part because of the population exodus, huge tracts of land are being managed by individuals or families. "We need more people returning to the land," he says. La Junquera, effectively abandoned for sixty years, has perhaps a dozen residents. Nearby Topares has a population of two hundred, with six children in the school. Everyone there says La Junquera is but a shadow of its past. The village goes back six centuries. Through much of its history it was a crossroads between the caliphate and the Christians, and absorbed the traditions of both.

Alfonso's family, I learn, has been farming here for five hundred years. Which helps explain his readiness to shift his commitments from business in Madrid to regenerative agriculture in the Altiplano. I express regret that I won't meet him on my visit. Jonathan tries to describe him. "Alfonso is like a Disney prince," he says. Rachel laughs and says, "I was expecting an older man with a cap and a sheepdog by his side."

———————

It is raw and windy—I've piled on layers, including my trusted red rain jacket—so when we go to camp our hosts are glad for our rental car. The camp vehicle, a mud-spattered white four-by-four, and Alfonso's pickup truck are not always available. To get to the camp area, crew staying in town often hike or bicycle. Today it is plenty muddy, and the dirt road is ribbed with ruts. Rachel and Jonathan jump out of the car when we arrive at the camp, a sparsely vegetated wedge of vale with a few small buildings. "These are our 'eco-porta-potties,'" Rachel says brightly. "Here are hay bales. Tunnels for plants. Yurts and bell tents." The latter are classic camping shelters held up with a single pole. She shrugs apologetically, perhaps self-conscious about this burst of enthusiasm, and concedes, "It doesn't look like much."

Jonathan points to a dip in the ground ringed by saplings: their firepit. "You can't have a camp without a campfire," he says. He bends down to show me how they're insulating the young trees. "We use sheep wool as mulch. It's a waste product here. And these ponds: We have a resource now—they didn't dry out this summer. Daniel Halsey made a design of swales and earthworks." Halsey, of Southwoods Ecosystems in Minnesota, is a permaculture designer and guiding consultant for ERC. "We wanted to come and immediately plant, but there was such a degree of compaction: solid, almost concrete below. So we used a subsoiler—a deep ripper—on contour to open up the soil. Otherwise the water sits there and evaporates."

He says they planted cover crops across the whole site the previous winter, though only one section received compost. "Rather than trying to do a wide area, we focused on one area," he says. When the cover crops grew tall, they faced the challenge of how to get the plant material incorporated into the soil—otherwise it would decay, oxidize, and block out sunlight for new growth. The solution turned out to be thirty-six hours of a thousand sheep, followed by rain.

As we walk around I learn the extent to which every detail has been thought through. The gray-water system irrigates newly planted trees. The swales, which collect water, serve as "mini wildlife corridors" while the plants they put in have multiple functions, including as windbreaks. "I wouldn't have guessed that our biggest challenge here would be winds," says Jonathan. "There are no native nitrogen-fixing trees here, so we're focusing on shrubs." He runs his hand over a bed of foliage the way someone might pat a beloved child's mop of hair. "These are mustards. Brassicas don't form mycorrhizal connections, so they establish quickly as pioneer plants." I hadn't known arugula fit into this category. No wonder I am seeing so much of it around here. The plants come to the scene in force to enhance soil conditions for higher-order vegetation.

We head over to the two-thousand-liter (528-gallon) water tank—an insurance policy for high summer—and Rachel shows us a system that monitors water level, humidity, and rainfall. This quantifies changes and also records a history to enable research. "This is an example of why the membership network is important," she says. "The member who installed this works in solar energy, and his company invested in this. We get a live feed every ten minutes. This is not something we would be able to do on our own. Often when we need something, just the right person with the right background shows up." Then again, she adds, often they don't. To a large extent serendipity has been driving what they can, or cannot, do at any given time.

On our way to lunch we pass the largest of the three ponds. Rachel says they're seeing lots of critters: frogs, water snakes, water rats, "loads of really nice dragonflies." Deer and wild boar come to drink, as do European wildcats. "If you come here in the summer it's teeming with life."

The kitchen cabin is under construction, and insulation for the walls is going in. The wooden awning that serves as a shelter extends from the back of the cabin. We squeeze around a large picnic table and I'm introduced to Rik, Ides, Dori, Frances, and Jo. That name I recognize: Jo Denham, who had helped me arrange the visit. There's a platter of bread and pots full of rice, curried vegetables, and lentils. Someone's mug reads I LOVE CAMPING, with a tent substituted for the usual heart.

The conversation settles on the Re-Generation Festival the camp hosted in September, a collaborative effort of Ecosia, AlVelAl, and locals that brought five hundred people to the site, three hundred of whom stayed over at camp and in the village. "It was so hot and sunny then," Jonathan says. Everyone nods, nostalgic for late summer and dragonfly season and impromptu music making.

One highlight was "planting to the beat": The Sound Matters team built a techno track from field recordings of spade work and frogs singing and, improbably, digitized soil monitoring data. They set up their console in the field as people planted rows of herbs and junipers, with occasional breaks for dancing. "That was such a joyful morning," Jo says. At night the main performance stage was set up in the village next to an illuminated sculpture of the beloved Iberian lynx. Volunteer staff included "Insecurity Guards" stationed by the bonfire who assisted in extinguishing self-doubt.

The rain starts while we're eating. For dessert we get a rainbow. Here on the Spanish high plains, these celestial apparitions have become almost routine. At the Re-Generation Festival, an alert photographer captured a shooting star. In the picture it looks like a slender teardrop as it sails through the sky.

Before we return to the village Ides Parmentier, a craftsman from Belgium, shows us the progress on the roundhouse. With weather conditions frequently harsh, having a permanent building on-site is important. (In Jonathan's words, the climate is "exactly what you want for six months of the year.") Jo and the others have been rebuilding the walls, stone by hefty stone, from the ruins of a centuries-old structure. Ides did the woodwork. There are weathered wooden posts and a reciprocal roof frame, sometimes called a Mandala roof, identified by its spiral geometry. The insulation is clay and straw. There's the wonderful warm smell of hay that's baked in summer sun.

Our last stop is the compost mound, a point of pride since fodder for decomposition is hard to come by. "We have difficulty to find enough composting materials," Ides says. "In places like Belgium and the Netherlands, they don't know what to do with all the cowshit!"

On Camp Altiplano's blog Jo Denham has an anecdote about the time they arranged to collect a bounty of pig manure several kilometers away but had no vehicle for transporting it. She and several others ended up bicycling to the farm and riding back, steep inclines and all, with burlap sacks of the stuff strapped to their backs. I recently caught a Facebook post in which a camp volunteer triumphantly displays on his palm an earthworm, a sign that there's finally some organic matter worth wriggling in.

Back in La Junquera we stop at Alfonso's farmhouse, where workmen are fixing the roof. They pause to let us through. Here we meet Alfonso's girlfriend, Yanniek Schoonhoven, who came to the area to write her master's thesis on AlVelAl and obstacles to regenerative agriculture. "I met Alfonso and basically never left," she says. "When I came here two years ago it was only me and Alfonso on the farm." Yanniek was one of the main drivers behind the Re-Generation Festival. She has also helped to launch Regeneration Academy, which offers training and research opportunities in land restoration. Several universities, including Yanniek's alma mater, Utrecht University, give credit for a semester's work. There is great demand among students for hands-on experience and research programs in regenerative agriculture, she says. The academy has rented two houses in Topares, which, she says, "increases the village population by 10 percent."

The ERC network is growing. March 2019 marked the beginning of Camp Vía Orgánica in Mexico, where the Regeneration International meeting took place a year and a half before. This camp isn't run by resident volunteers; instead twenty-five people paid a modest fee for a ten-day restoration experience complete with training activities, organic meals, and accommodations in proper adobe cabins. "People are now bringing these skills to where they live. Four were farmers who have their own land, including two in Mexico, one in the far north where it is very dry," says Business Development Coordinator Ashleigh Brown. She says a participant from France is in the process of setting up a camp on his land. They saw that the model of brief learning sessions on existing sites worked well.

Camp Chocaya in the Bolivian Andes was open from October 2019 to January 2020. In a collaboration with Chocaya village and local partners, this camp focused on designing and implementing an agroforestry restoration plan. One goal was to provide a model for neighboring communities also contending with hillside erosion and dry-season droughts. Camp Uthai in Thailand, which welcomed its first campers in October 2019, is creating a food forest on twenty-three hectares (fifty-seven acres) of former rice paddy. In California, Camp Paradise was established as an "ecosystem restoration disaster response" camp, supporting the Butte County community that was devastated by the 2018 Camp Fire to rehabilitate the burned area, so that it is more productive and resilient to fire.

Meanwhile, Camp Altiplano has evolved, pivoting and correcting course as they learn and move forward. Rather than having volunteers year-round, the camp will host people at certain times for specific projects. They plan to hire a local person to coordinate projects and operations. Ashleigh says this will "professionalize it and give it more structure. The project has matured after [our] learning lots of lessons and experiencing what did and didn't work. It will be less like 'woofing' and more like a set program."[15]

Among the lessons learned, Ashleigh says, is that "building something from scratch is really hard." Hence the new focus on locations that have infrastructure. Now organizers can devote more time to expanding the number of camps. The challenge, she says, is funding. The organization has a goal of one hundred camps in various parts of the world by 2030. In order to get there, they would need two thousand members; they currently have fewer than one thousand.

Says Ashleigh: "There is a lot of interest from many, many countries. Until we expand our team it will be harder to keep on this trajectory even though that's what needs to happen. At the moment I'm doing everything, basically. We have an amazing team of volunteers. With our current income what we're doing is really cool. But it's really knackering for me, and quite lonely sometimes." There is no shortage of people who want to work, have land, or have expertise, she says. It's a matter of being able to mobilize this.

We regroup at the farmhouse kitchen to say goodbye. We've got that sketchy drive back to Vélez Blanco ahead of us, unlikely to get any easier once dusk falls. I join Jonathan at his computer to suggest reading sources, and we chat for a few moments. With the ecological and climate crises, "people are going through a kind of grief," he says. "We've all gone through periods of huge depression for what's going on." He says that Camp Altiplano and efforts like it can provide hope, community, and a sense of meaning that may be absent in other realms, particularly those in which the reality of crisis is denied. And that taking action is an antidote to depression.

I recall something Yanniek had said just a few minutes before, that some of the young people who find their way to the Regeneration Academy are "close to giving up on the world" and that the curriculum offers "an example of what's possible." I think of the bright faces I'd walked past as Yanniek led us through the study space on our way back to the farmhouse, young people focused on their projects. For all the world they looked like contented scholars, engaged with their work and with ripe futures before them. I am reminded that when you're in your twenties, your countenance doesn't necessarily betray your pain. I was likely the same way.

I feel a new kind of grief, not for myself but for today's youth, those sensitive to nature and who never had the chance to take forests, birds, and insects for granted the way I was able to. This entire cohort and those to follow are deprived of the incalculable solace of predictable seasons, the transition from, say, winter to spring so defined and familiar you know it in your bones. If I could change that, and somehow give them the blithe assurance of nature's permanence that I'd enjoyed at their age, I would do it in a heartbeat.

I think of what I've learned from reporting this book. Not just about restoring ecosystems, but about the power in focusing on what's possible rather than impediments; on best possible outcomes instead of what one dreads; on action rather than inertia. It occurs to me that the Ecosystem Restoration Camp movement is in many ways a manifestation of best possible outcome thinking.

Let's go back to that pivotal consensus question: Given that the goal—in this case, to restore earth's degraded ecosystems—is impossible, what would you do if it were achievable?

How might someone answer this? Maybe, to start, find land to restore and live there, or restore the land you're living on. Build soil organic matter. Keep water in the ground. Grow some healthy food. Establish trees and other deep-rooted perennial plants. Bring animals onto the land. Gather with others who share your passion for healing the earth. Ditch the apathy. Jettison the cynicism. Make music and art. Dance. Create new things and refurbish old things in beautiful ways. Keep learning how to do things better and keep records of progress. Expect unforeseen challenges and find ways to laugh at them. Be resourceful and open to offers of help. Celebrate small victories. Spread the word.

This is what the folks at Camp Altiplano are doing. Accepting the inevitability of feared outcomes, which frequently takes the form of ironic distance, serves to absolve us of responsibility. These young people are taking responsibility. They are living in line with their best possible outcomes. In doing so they are ever-so-slightly tilting the universe in that direction. Who knows how powerful these actions will be, and who will be inspired by them.

"I feel like we're in the anteroom before the stage curtain, that we're at stage left," Jonathan says. "This work is about to be thrust into the spotlight. At some point people will say, 'We need our rivers.'"

And the global task of ecosystem restoration will begin in earnest. Once ideas on how to achieve what's assumed to be impossible are articulated, that goal is no longer impossible.

Let's get inspired.

Acknowledgments

T his book came to fruition thanks to the many people who have provided help and been by my side through the process of research and writing, and for this I offer my gratitude. I want to thank everyone who invited me into their lives in New Mexico, Norway, Hawai'i, eastern Washington, and Spain, as well as those who shared their experience and insights from afar. Thanks, too, are due to Diana Donlon and Alexandra Groome Klement, who heroically read drafts and offered critiques, usually on deadline. I want to note eco-restoration thinkers and practitioners I interact with in conferences and meetings and on social media, who often spark new ideas and meaningful discussions. I also want to express appreciation to the growing regenerative agriculture community in my region, and to my friends and teachers at the karate dojo, who have helped me discover strength I didn't know I had (and give me a chance to let off steam!)

The Globetrotter Foundation provided financial support for reporting in Spain. Premiere 1 Supplies helped set me up with fencing for sheep; while our summerlong "sheepover" didn't make it into the book, I treasure the months with these sweet animals. Many thanks to the team at Chelsea Green for taking another book journey with me, especially to my editor, Brianne Goodspeed, for keeping me in line. Again and again my agent, Laura Gross, proves how indispensable she is to me, and I am endlessly grateful. And as always I want to thank my family, for providing perspective, for humoring me, and for their love.

Notes

Introduction

1. John Todd, *Healing Earth* (Berkeley, CA: North Atlantic Books, 2019), 3.
2. "The Biggest Environmental Challenges of 2017," Nature Conservancy, accessed January 22, 2020, https://www.nature.org/en-us/what-we-do/our-insights/perspectives/the-biggest-environmental-challenges-of-2017.
3. "Main Street Project: What We Do," Main Street Project, accessed January 22, 2020, https://mainstreetproject.org/what-we-do.
4. Peter Donovan, "Atlas of Biological Work: growing a shared intelligence a shared politics and economics on soil health and watershed function…and maybe more…" Soil Carbon Coalition, version May 2019, p. 10, soilcarboncoalition.org/files/atlasbooklet.pdf.
5. "Leopold's Game Management," University of New Mexico Searchable Ornithological Research Archive, accessed January 22, 2020, https://sora.unm.edu/sites/default/files/journals/auk/v050n03/p0376-p0377.pdf.
6. Peter Donovan, "Atlas of Biological Work," p. 5, Soil Carbon Coalition, https://soilcarboncoalition.org/files/atlas.pdf.

Chapter 1. The Great Work of Our Time

1. Alexandra Groome, "Meet John D. Liu, the Indiana Jones of Landscape Restoration," March 7, 2016, accessed January 22, 2020, https://regenerationinternational.org/2016/03/07/meet-john-d-liu-the-indiana-jones-of-landscape-restoration.
2. "Let's #StopSoilErosion to Ensure a Food Secure Future," Food and Agriculture Organization of the United Nations, May 15, 2019, accessed January 22, 2020, http://www.fao.org/fao-stories/article/en/c/1192794.
3. Peter Donovan, "The Soil Carbon Challenge," accessed January 22, 2020, https://soilcarboncoalition.org/challenge.
4. "Restoring China's Loess Plateau," World Bank, March 15, 2007, accessed January 22, 2020, http://www.worldbank.org/en/news/feature/2007/03/15/restoring-chinas-loess-plateau.

5. Soil organic matter is approximately 58 percent carbon.
6. John D. Liu, "Back to the Garden," Green the Sinai, May 2019, accessed January 22, 2020, https://www.greenthesinai.com/back-to-the-garden -john-d-liu?fbclid=IwAR2gUP6tNIRqcEa9xfBVWv3DdlrdAbngNsF DCVAuR2isJanCXXflp98uYKA.
7. According to the USDA, every percentage point increase in soil organic carbon represents an additional twenty thousand gallons (or more) of water per acre held on the land: https://www.nrcs.usda.gov/Internet /FSE_DOCUMENTS/stelprdb1082147.pdf.
8. Chris Winter, "Raj Patel on How to Break Away from Capitalism," *Yes! Magazine* online, October 23, 2018, accessed January 22, 2020, http:// www.yesmagazine.org/new-economy/raj-patel-on-how-to-break -away-from-capitalism-20171023. (The quote has also been attributed to Fredric Jameson and Slavoj Žižek.)
9. "Impact of Dams on the People of Mali," p. 4, Wetlands International, accessed January 22, 2020, http://archive.wetlands.org/Portals/0 /publications/Count%20Form/Brochure/WI%20Upper%20Niger%20 brochure.pdf.
10. David Graeber, "Why Capitalism Creates Pointless Jobs," *Evonomics*, accessed January 22, 2020, http://evonomics.com/why-capitalism -creates-pointless-jobs-david-graeber.
11. Here's a column I wrote on this topic: https://www.theguardian.com /commentisfree/2017/apr/03/climate-change-water-fossil-fuel.

Chapter 2. Life Begets Life

1. "Frequently Asked Questions: Is Desertification a Global Problem?," United Nations Convention to Combat Desertification, accessed January 23, 2020, https://www.unccd.int/frequently-asked -questions-faq.
2. Elisabet Sahtouris, "Prologue to a New Model of a Living Universe," in *Mind Over Matter*, p. 14, 2007, accessed January 23, 2020, http://www .sahtouris.com/pdfs/MindBeforeMatterChapter.pdf.
3. "Desert," National Geographic Resource Library, accessed January 23, 2020, https://www.nationalgeographic.org/encyclopedia/desert.
4. "Greening the Desert," 2007, YouTube, accessed January 23, 2020, https://www.youtube.com/watch?v=sohI6vnWZmk&t=3s.
5. Nathan Halverson, "What California Can Learn from Saudi Arabia's Water Mystery," *Reveal*, Center for Investigative Reporting, April 27,

2015, accessed January 23, 2020, https://www.revealnews.org/article/what-california-can-learn-from-saudi-arabias-water-mystery.

6. "Saudi Arabia's Great Thirst," *National Geographic*, accessed January 23, 2020, https://www.nationalgeographic.com/environment/freshwater/saudi-arabia-water-use.

7. J. S. Famiglietti, "The Global Groundwater Crisis," *Nature Climate Change* 4 (November 2014), 945–48, accessed January 23, 2020, https://doi.org/10.1038/nclimate2425.

8. Tom Dale and Vernon Gill Carter, *Topsoil and Civilization*, rev. ed. (Norman: University of Oklahoma Press, 1974), 6.

9. Dale and Carter, *Topsoil and Civilization*, 240.

10. J. Russell Smith, *Tree Crops: A Permanent Agriculture* (New York: Harcourt, Brace, 1929), Journey to Forever Farm Library, chapter 1: "How Long Can We Last?," accessed January 23, 2020, http://journeytoforever.org/farm_library/smith/treecrops1.html#ch1.

11. The term *wadi* refers to a valley or streambed that, though mostly dry, may run with water during the rainy season.

12. Geoffrey Lawton, Swale and Crop Layout, Permaculture Research Institute, August 2010, accessed January 23, 2020, https://permaculturenews.org/2010/08/06/letters-from-jordan-on-consultation-at-jordans-largest-farm-and-contemplating-transition.

13. Qur'an, 18.32, accessed January 23, 2020, https://quran.com/18.

14. https://www.nature.com/articles/s41477-018-0205-y?WT.feed_name=subjects_evolution.

15. S. Yoshi Maezumi et al., "The Legacy of 4,500 Years of Polyculture Agroforestry in the Eastern Amazon," *Nature Plants* 4 (August 2018), 540–47, accessed January 23, 2020, https://doi.org/10.1038/s41477-018-0205-y.

16. "How to Make It Rain in the Desert," *Earth Repair Radio with Andrew Millison*, Episode 002, accessed January 23, 2020, https://www.earthrepairradio.com.

17. Jim Robbins, "The Climate Connection: Unraveling the Surprising Ecology of Dust," *Yale Environment 360*, November 30, 2017, https://e360.yale.edu/features/climate-connection-unraveling-the-surprising-ecology-of-dust.

18. Judith D. Schwartz, "Botanist Brings Trees to the Israeli Desert," *Pacific Standard*, July 21, 2011, accessed January 23, 2020, https://psmag.com/environment/botanist-brings-trees-to-the-israeli-desert-33935.

19. T. J. Lyons, "Clouds Prefer Native Vegetation," *Meteorology and Atmospheric Physics* 80 (June 2002), 131–40, https://doi.org/10.1007/s007030200020.

20. Douglas Sheil, "Forests Versus Hurricanes," *Forests News*, Center for International Forestry (CIFOR), September 18, 2017, accessed January 23, 2020, https://forestsnews.cifor.org/51566/forests-versus-hurricanes?fnl=en.

21. "Ecological Rainfall Infrastructure: Investment in Trees for Sustainable Development," WeForest, January 15, 2016, accessed January 23, 2020, http://www.weforest.org/newsroom/ecological-rainfall-infrastructure-investment-trees-sustainable-development.

22. Millán M. Millán, "Land Use Changes and Their Impacts on Extreme Events," presentation to European Commission Conference on Land as a Resource, Brussels, June 19, 2014, accessed January 23, 2020, http://ec.europa.eu/environment/land_use/pdf/millanslides.pdf.

23. "Over a Half Million Corals Destroyed by Port of Miami Dredging, Study Finds," *Science Daily*, May 30, 2019, accessed January 23, 2020, https://www.sciencedaily.com/releases/2019/05/190530141448.htm; Miranda Fox, "Dredging Florida's Corals to Death," Earth Justice, May 27, 2016, accessed January 23, 2020, https://earthjustice.org/blog/2016-may/dredging-florida-s-corals-to-death.

24. "2018 Right Livelihood Award Laureates Announced," Right Livelihood Foundation, September 28, 2018, accessed January 23, 2020, https://www.rightlivelihoodaward.org/media/2018-right-livelihood-award-laureates-announced.

Chapter 3. Beyond the Impossible

1. Names of people in the Cabezon community have been changed for privacy.

2. "Soul Biographies: Consensus," Nic Askew, accessed January 23, 2020, http://nicaskew.com/collection/consensus.

Chapter 4. They Belong to This Land

1. Richard Martyn-Hemphill, "In Norway, Fighting the Culling of Reindeer with a Macabre Display," *New York Times*, December 6, 2017, accessed January 25, 2020, https://www.nytimes.com/2017/12/06/world/europe/reindeer-norway-trial.html.

2. Sápmi is the region where Sámi people live, most of which is above the Arctic Circle. It used to be called Lapland, which many now consider a derogatory term.

3. From an interview with Máret Ánne Sara during an event at Documenta 14, an exhibit of international art in 2017 that featured her work. Accessed January 23, 2020, https://www.documenta14.de/en/artists /13491/maret-anne-sara.

4. "Máret Ánne Sara: l'art au service people Same," accessed January 23, 2020, https://vimeo.com/306575992.

5. Documenta 14 interview.

6. This has been true through many drought years. Unfortunately, in 2019 drought was so severe that some parts of the river stopped flowing during the dry season.

7. Mariska te Beest et al., "Reindeer Grazing Increases Summer Albedo by Reducing Shrub Abundance in Arctic Tundra," *Environmental Research Letters* 11, no. 12 (December 22, 2016), accessed January 23, 2020, https://iopscience.iop.org/article/10.1088/1748-9326/aa5128.

8. Jukka Käykhö and Tim Horstkotte, "Reindeer Husbandry Under Global Change in the Tundra Region of Northern Fennoscandia," University of Turku, March 1, 2017, DOI: 10.13140/RG.2.2.22151.39841.

9. "Pleistocene Park: Restore High Productive Grazing Ecosystems in the Arctic and Mitigate Climate Change," accessed January 23, 2020, https://pleistocenepark.ru.

10. Adam Wernick, "A Bold Plan to Slow the Melt of Arctic Permafrost Could Help Reverse Global Warming," *PRI's Living on Earth*, April 11, 2017, accessed January 23, 2020, https://www.pri.org/stories /2017-04-30/bold-plan-slow-melt-arctic-permafrost-could-help -reverse-global-warming.

11. Yadvinder Malhi et al., "Megafauna and Ecosystem Function from the Pleistocene to the Anthropocene," *Proceedings of the National Academy of Sciences* 113, no. 4 (January 26, 2016), 838–46, accessed January 23, 2020, https://doi.org/10.1073/pnas.1502540113.

12. Maria Väisänen et al., "Long-Term Reindeer Grazing Limits Warming-Induced Increases in CO_2 Released by Tundra Heath Soil: Potential Role of Soil C Quality," *Environmental Research Letters* 10, no. 9 (September 15, 2016), accessed January 23, 2020, https://iopscience .iop.org/article/10.1088/1748-9326/10/9/094020.

13. Johann Olafsson et al., "Effects of Summer Grazing by Reindeer on Composition of Vegetation, Productivity and Nitrogen Cycling," *Ecography*, January 19, 2009, accessed January 23, 2020, https://doi .org/10.1034/j.1600-0587.2001.240103.x.

14. Jinny McCormick, "This Reindeer Battalion of WWII Was Braver than Soviets, Tougher than Tanks," War History Online, August 18, 2016, accessed January 23, 2020, https://www.warhistoryonline.com/world -war-ii/reindeer-battalion-wwii-braver-soviets-tougher_tanks.html.

15. "Atlantic Ocean Road," Dangerous Roads, accessed January 23, 2020, https://www.dangerousroads.org/europe/norway/164-atlantic-ocean -road-norway.html.

16. Marie Roué, "Sami Knowledge in a Changing World—Sami Ecology and Science of Snow," Kunsthall Trondheim, May 2017, accessed January 23, 2020, https://vimeo.com/218131444.

17. Roué, "Sami Knowledge in a Changing World."

18. Ánde Somby, "When a Predator Culture Meets a Prey Culture," Kunsthall Trondheim, May 2017, accessed January 23, 2020, https:// vimeo.com/217594529.

19. Tom Batchelor, "Norway Approves Copper Mine in Arctic Described as 'Most Environmentally Damaging Project in Country's History,'" *U.K. Independent*, February 14, 2019, accessed January 23, 2020, https://www.independent.co.uk/news/world/europe/norway -copper-mining-arctic-finnmark-pollution-environment-damage -fjords-a8778891.html.

20. "Norwegianization of the Sami," Teach Indigenous History, accessed January 23, 2020, https://teachik.com/norwegianization-of-the-sami.

21. Somby, "When a Predator Culture Meets a Prey Culture."

22. Hugo Reinert, "The Skulls and the Dancing Pig: Notes on Apocalyptic Violence," *Terrain: Anthropologie & Sciences Humaines* 71 (April 2019), accessed January 23, 2020, DOI: 10.4000/terrain.18051.

23. Tor A. Benjaminsen, Hugo Reinert, Espen Sjaastad, and Mikkel Nils Sara, "Misreading the Arctic Landscape: A Political Ecology of Reindeer, Carrying Capacities, and Overstocking in Finnmark, Norway," *Norsk Geografisk Tidsskrift—Norwegian Journal of Geography* 69, no. 4 (June 2015), https://doi.org/10.1080/00291951.2015.1031274.

24. http://hugo.re/wp-content/uploads/2019/05/Reinert-The-Skulls-and -the-Dancing-Pig.pdf.

25. Megafauna and Ecosystem Function: From the Pleistocene to the Anthropocene, conference at the University of Oxford College of St. John, March 18–20, 2014, https://oxfordmegafauna.weebly.com.

26. Malhi et al., "Megafauna and Ecosystem Function."

27. Malhi et al., "Megafauna and Ecosystem Function."

28. Willow Alexandria Blough, "8 Unexpected Uses for Elephant Dung," Africa Geographic, November 6, 2015, accessed January 23, 2020, https:// africageographic.com/blog/8-unexpected-uses-for-elephant-dung.

29. Malhi et al., "Megafauna and Ecosystem Function."

30. Joshua E. Brown, "Whales as Ecosystem Engineers," *UVM Today*, accessed January 23, 2020, https://www.uvm.edu/uvmnews/news /whales-ecosystem-engineers.

31. Malhi et al., "Megafauna and Ecosystem Function."

32. Malhi et al., "Megafauna and Ecosystem Function."

33. I wrote about Henggeler in *Water in Plain Sight: Hope for a Thirsty World*.

34. Matt Brann, "Wild Donkeys in Kimberley Scientific Trial in Aerial Sights of Shooters," *ABC Rural*, May 22, 2018, accessed January 23, 2020, https://www.abc.net.au/news/rural/2018-05-23/kachana-wild -donkey-research-in-the-sights-of-aerial-shooters/9777062.

35. "Feral Donkey," Government of Western Australia Department of Primary Industries and Regional Development, December 19, 2014, accessed January 23, 2020, https://www.agric.wa.gov.au/pest-mammals /feral-donkey.

36. "Fate of the Wild Donkeys on Kachana?" Kachana, May 2018, accessed January 23, 2020, http://www.kachana.com/Fate%20of%20the%20wild %20donkeys%20of%20Kachana.pdf.

37. Chris Henggeler, "Wild Donkey Project," Kachana Station, November 2017, accessed January 23, 2020, https://www.kachana-station.com /projects/wild-donkey-project.

38. Brann, "Wild Donkeys in Kimberley Scientific Trial."

39. Brann, "Wild Donkeys in Kimberley Scientific Trial."

40. Dingo for Biodiversity Project, University of Technology Sydney: Centre for Compassionate Conservation, accessed January 23, 2020, https://www.dingobiodiversity.com.

41. Centre for Compassionate Conservation, University of Technology Sydney, accessed January 23, 2020, https://www.uts.edu.au/research -and-teaching/our-research/centre-compassionate-conservation.

42. Arian Wallach, "Feminist Ferals," Political Ecology Research Centre, Massey University, University of New Zealand, accessed January 23, 2020, http://perc.ac.nz/wordpress/feminist-ferals.

43. "Feral Horse (*Equus caballus*) and Feral Donkey (*Equus asinus*)," Australian Government Department of Sustainability, Environment, Water, Population and Communities, accessed January 23, 2020, http://www.environment.gov.au/system/files/resources/b32a088c -cd31-4b24-8a7c-70e1880508b5/files/feral-horse.pdf.

44. Arian Wallach et al., "Invisible Megafauna," *Conservation Biology* 32, no. 4 (April 2018), 962–65, accessed January 23, 2020, DOI: 10.1111 /cobi.13116.

45. Xavier La Canna, "Australia's Introduced Animals: Eradication Programs Under the Spotlight," Australian Broadcasting Corporation, September 12, 2017, accessed January 23, 2020, https://mobile.abc.net .au/news/2017-09-12/should-australia-rethink-eradication-programs -of-feral-animals/8830998?pfmredir=sm.

46. Patrick Barkham, "Dutch Rewilding Experiment Sparks Backlash as Thousands of Animals Starve," *U.K. Guardian*, September 27, 2018, accessed January 23, 2020, https://www.theguardian.com/environment /2018/apr/27dutch-rewilding-experiment-backfires-as-thousands-of -animals-starve.

47. Isabella Tree, *Wilding: The Return of Nature to a British Farm* (London: Picador, 2018), 43.

48. Isabella Tree @5X15, May 2018, accessed January 23, 2020, https:// www.youtube.com/watch?v=4cug0KcTnXI.

49. Isabella Tree @5X15.

Chapter 5. From Scarcity to Abundance

1. Michelle Broder Van Dyke, "A New Hawaiian Renaissance: How a Telescope Protest Became a Movement," *U.K. Guardian*, August 17, 2019, accessed January 24, 2020, https://www.theguardian.com /us-news/2019/aug/16/hawaii-telescope-protest-mauna-kea.

2. "Introduced Species in Hawaii," Earlham College Senior Seminar 2002, accessed January 24, 2020, http://legacy.earlham.edu/~meckehe /introduction.htm.

3. "Hawaiian Land System," Hawaiian Kingdom, accessed January 24, 2020, https://www.hawaiiankingdom.org/land-system.shtml.

4. Biodiversity for a Livable Climate, https://bio4climate.org.

5. Deborah Caulfield Rybak, "Who Are These Guys? Enthusiasm Over the New Owners of A&B's Agricultural Lands Runs High, Though Almost Nothing Is Known About Them," *MauiTime*, December 30, 2018, accessed January 24, 2020, https://mauitime.com/news/business/who-are-these-guys-enthusiasm-over-the-new-owners-of-abs-agricultural-lands-runs-high-though-almost-nothing-is-known-about-them.

6. Garth Staply, "Oakdale Orchard Dispute Fetches $4 Million from Almond Giants," *Modesto Bee*, February 1, 2014, accessed January 24, 2020, https://www.modbee.com/news/local/oakdale/article3160103.html.

7. Deborah Caulfield Rybak, "The Man with the Plan: Mahi Pono's New General Manager Larry Nixon Wants More Bees and Bugs and Less Corporate Hierarchy as He Rehabilitates and Replants Central Maui's Cropland," *MauiTime*, February 27, 2019, accessed January 24, 2020, https://mauitime.com/news/business/the-man-with-the-plan-mahi-ponos-new-general-manager-larry-nixon-wants-more-bees-and-bugs-and-less-corporate-hierarchy-as-he-rehabilitates-and-replants-central-mauis-cropland.

8. "A Destructive 2018 Hurricane Season Comes to a Quiet End," *Hawaii News Now*, November 29, 2018, http://www.hawaiinewsnow.com/2018/11/30/destructive-hawaii-hurricane-season-comes-quiet-end.

9. "Thank you" in the Hawaiian language.

10. Chip Fletcher, "IUCN: We Need Public Service Announcements About Climate Change," *Honolulu Civic Beat*, September 6, 2016, accessed January 24, 2020, https://www.civilbeat.org/2016/09/iucn-we-need-public-service-announcements-about-climate-change.

11. "Timeline: East Maui Streams," Maui Sierra Club, accessed January 24, 2020, http://mauisierraclub.org/wp-content/uploads/2017/02/Timeline_EastMauiStreams2017.pdf.

12. Anita Hofschneider, "This Native Hawaiian Taro Farmer Has Been Fighting A&B for Decades," *Honolulu Civic Beat*, May 2, 2019, accessed January 24, 2020, https://www.civilbeat.org/2019/05/this-native-hawaiian-taro-farmer-has-been-fighting-ab-for-decades.

13. "Maui GMO Moratorium," Organic Hawaii, accessed January 24, 2020, https://organichawaii.org/maui-gmo-ban/?v=7516fd43adaa#.XL-NyetKiQM.

14. Robynne Boyd, "Genetically Modified Hawaii," *Scientific American*, December 8, 2008, accessed January 24, 2020, https://www.scientificamerican.com/article/genetically-modified-hawaii.

15. Jeri Di Pietro and Mickey, "Report Released: Syngenta Incident at Waimea School," Island Breath, January 12, 2007, accessed January 24, 2020, http://www.islandbreath.org/2006Year/16-farming/0616-20 WaimeaPoison.html; Christopher Pala, "Pesticides in Paradise: Hawaii's Spike in Birth Defects Puts Focus on GM Crops," *U.K. Guardian*, August 23, 2015, accessed January 24, 2020, https://www .theguardian.com/us-news/2015/aug/23/hawaii-birth-defects -pesticides-gmo.

16. Paul Kolberstein, "GMO Companies Are Dousing Hawaiian Island with Toxic Pesticides," *Grist*, June 16, 2014, accessed January 24, 2020, https://grist.org/business-technology/gmo-companies-are-dousing -hawaiian-island-with-toxic-pesticides.

17. "Profile: Lorrin Pang, MD," Maui GMO Moratorium News, accessed January 24, 2020, https://www.mauigmomoratoriumnews.org/profile /lorrin-pang.

18. Poi, a starch made from boiled taro root, is a traditional Hawai'i staple food.

19. https://www.hawaii.edu/uhwo/clear/home/lawaloha.html.

20. "Mongoose," Hawaii Invasive Species Council, accessed January 24, 2020, http://dlnr.hawaii.gov/hisc/info/invasive-species-profiles/mongoose.

21. "Video: The Power of Bamboo," Permaculture Research Institute, accessed January 24, 2020, https://permaculturenews.org/2014/01/11 /power-bamboo-video.

22. He is also author of *Restoring the Soil: A Guide for Using Green Manure/ Cover Crops to Improve the Food Security of Smallholder Farmers*, https:// foodgrainsbank.ca/uploads/Restoring%20the%20Soil.pdf.

23. Judith D. Schwartz, "Soil Called Crucial to Combating Climate Change," *FERN's Ag Insider*, May 15, 2017, accessed January 24, 2020, https:// thefern.org/ag_insider/soils-called-crucial-combating-climate-change.

24. "Video: Life in Syntropy," Agenda Gotsch, December 2, 2015, accessed January 24, 2020, https://www.youtube.com/watch?v=gSPNRu4ZPvE.

25. "Forestry," Hōkūnui, accessed January 24, 2020, https://www.hokunui .com/forestry.

26. Kehaulani Cerizo, "'Accidental' Fire Sparks Fear, $1M in MECO Damage," August 3, 2019, accessed January 24, 2020, https://www .mauinews.com/news/local-news/2019/08/accidental-fire-below -pukalani-sparks-fear-1m-in-meco-damage.

27. Wendy Osher, "Mahi Pono's First Planting Signals Start to 'Re-Greening' Maui," *Maui Now*, August 23, 2019, accessed January

24, 2020, https://mauinow.com/2019/08/23/mahi-ponos-first
-planting-signals-start-to-re-greening-maui.

28. Deborah Caulfield Rybak, "Mahi Pono Is Coming for Our Water,"
    MauiTime, October 30, 2019, accessed January 24, 2020, https://
    mauitime.com/news/politics/mahi-pono-is-coming-for-our-water.

Chapter 6. Busting the Myth

1. "#WomenInAg," US Department of Agriculture, accessed January 25,
   2020, https://www.usda.gov/sites/default/files/documents/United
   -States-Womeninag.pdf.

2. "About Women, Food and Agriculture Network," Women Food & Ag
   Network, accessed January 25, 2020, http://www.wfan.org/about.

3. Liz Forster, "Colorado Women in Ranching: A Spirit of Nurturing
   Sustainability Is Alive at San Juan Ranch," *Colorado Springs Gazette*,
   December 21, 2018, accessed January 25, 2020, https://gazette.com
   /news/colorado women-in-ranching-a-spirit-of-nurturing-sustainability
   -is/article_69b1b238-0564-11e9-b067-d/ef135f22f3.html.

4. Amanda Lucier, "Female Ranchers Are Reclaiming the American West,"
   New York Times, January 11, 2019, accessed January 25, 2020, https://
   www.nytimes.com/2019/01/11/business/women-ranchers-american
   -west-photo-essay.html.

5. "Women and Decent Work," Food and Agriculture Organization of the
   United Nations, accessed January 25, 2020, http://www.fao.org
   /rural-employment/work-areas/women-and-decent-work/en.

6. "Women and Girls: Women Smallholders," Project Drawdown, accessed
   January 25, 2020, https://www.drawdown.org/solutions/women
   -and-girls/women-smallholders.

7. Kun Li, "Malnutrition, Disease and Food Insecurity—a Perfect Recipe
   for Disaster for Somalia's Children," UNICEF Somalia, June 12, 2017,
   accessed January 25, 2020, https://www.unicef.org/somalia/pp_20083.html.

8. "Video: Resilience in Practice," Fibershed, November 13, 2018,
   accessed January 25, 2020, https://www.youtube.com/watch?v
   -1wKTDUEBDf8.

9. "Video: BCB Shepherdess 2012," Brittany Cole Bush, accessed
   January 25, 2020, http://brittanycolebush.com.

10. "#AftertheFire Webinar," Regrarians Ltd., January 7, 2018, accessed
    January 25, 2020, https://www.youtube.com/watch?v=26sW_y0a9n0.

11. "Resilience in Practice."

12. "BCB Shepherdess 2012."

Chapter 7. Science and Inspiration

1. George Orwell, *Homage to Catalonia* (New York and London: Harcourt Brace Jovanovich, 1980 edition), 30.
2. Willem Van Cotthem, "Desertification in Spain," *Desertification* (blog), October 30, 2016, accessed January 25, 2020, https://desertification .wordpress.com/2016/10/30/desertification-in-spain.
3. The full line, Millán points out, is "Water begets water, soil is the womb, and vegetation the midwife."
4. Millán Millán, "Land Use Changes in the Mediterranean May Be Triggering Large Weather Shifts," *Soil and Water: A Larger-Scale Perspective* 52 (November 2015), European Commission Center for Environment Policy, 7.
5. The first feedback is the loss of transpiration from vegetation; the second is loss of soil.
6. "The Importance of Wetlands," Ramsar Convention, accessed January 25, 2020, https://www.ramsar.org/about/the-importance-of-wetlands.
7. Study of Critical Environmental Problems (SCEP), *Man's Impact on the Global Environment* (Cambridge, MA: MIT Press), 1970, 52.
8. SCEP, *Man's Impact on the Global Environment*, 126.
9. "Landscapes First," Landscape Finance Lab, accessed January 25, 2020, http://www.landscapefinancelab.org; "Land to Market," Savory Global, accessed January 25, 2020, https://www.savory.global/land-to-market; "Realigning the Economics of Agriculture," Regen Network, accessed January 25, 2020, https://www.regen.network.
10. "Theory U," Otto Scharmer, accessed January 25, 2020, http://www .ottoscharmer.com/theoryu.
11. "Video: 4 Returns in Spain," Commonland, January 4, 2016, accessed January 25, 2020, https://www.youtube.com/watch?v=xbEuJ2RorMw &feature=youtu.be.
12. Tom Philpott, "Holy Shit! Almonds Require a Ton of Bees," *Mother Jones*, May 25, 2015, accessed January 25, 2020, https://www.mother jones.com/food/2015/05/almonds-now-require-85-percent-us-beehives.
13. The subtitle of *Small Is Beautiful* is *Economics as if People Mattered*.
14. E. F. Schumacher, *Small Is Beautiful* (New York: Harper & Row, 1973, First Perennial Edition, 1975), 15.
15. Stands for World Wide Opportunities on Organic Farms.

Index

About the Author

Tony Eprile

Judith D. Schwartz is a journalist whose work explores nature-based solutions to global environmental and economic challenges. She writes on this theme for numerous publications and speaks at venues around the world. She is the author of *Cows Save the Planet* and *Water in Plain Sight*. A graduate of the Columbia Journalism School and Brown University, she lives in southern Vermont.